I0504031

Industry 4.0:

Navigating the Manufacturing Revolution in ASEAN

Michael Deng & Colin Koh

About the Authors

Michael Deng (BSc. Tianjin University, MBA. NUS)

 Michael is a veteran business executive with more than 28 years of sales management experience across the Asia Pacific region, with special focus on the manufacturing, power & utility, smart city and infrastructure markets. Over his long career, Michael led business development, sales and general management for many renowned multi-national corporations, including General Electric, Oracle, Invensys, Sybase, and Autodesk.

As the pioneer of digital business for Asia, Michael has overseen numerous major commercialised projects that draw on his extensive combined OT and IT experience. He also leads smart infrastructure projects with regional governments and major industrial organizations. His deep understanding of industry eco-systems and market trends throughout Asia, allows Michael to guide businesses and governments along their digital transformation journeys.

Michael is an IEEE and Power & Energy Society member, widely involved in driving industry 4.0 initiatives across Asia-Pacific as a whole. He graduated from Tianjin University in 1994, with a computer engineering & science degree, and then achieved an MBA from the National University of Singapore in 1999.

Colin Koh (BBA. RMIT)

 Colin has more than 30 years of experience pioneering the latest technology within the ASEAN region, involved in the Industrial IoT/M2M, automation, 3D printing, connectivity, energy & environmental business since the mid-'80s. Colin provides mentorship and advisory services to the industry, especially focused on regional SMEs implementing digital transformation in line with the emerging Industry 4.0 concepts. He is a certified IoT specialist, certified Digital Transformation Management professional, and alumni of MIT Sloan Scholl of Management Executive program in Artificial Intelligence and IoT. Attained Singapore National Infocomm Competency Framework (NICF) in Enterprise Architecture. Competency of Smart Industry Readiness Index (SIRI) implementation.

Colin held the role of President of the Singapore Industrial Automation Association from 2001-2005 and was Vice-Chair of Industry Committee of Energy Efficiency under the Sustainable Energy Association of Singapore, sub-committee under the National Climate Change Committee in Singapore. Current member of IoT Technical Committee of IT Standards Committee (ITSC). He is also a member of International Society of Automation (ISA), the Singapore Water Association (SWA), and a former committee member of IoT Asia Conference and Exhibition.

Colin is a technology evangelist and driving force of Industry 4.0 in the ASEAN bloc. Delivering dozens of insightful industry papers and influential speeches in the regional conferences and business forums. He was featured in Channel News Asia (CNA) Start-Up series 6 as an industry expert. Member of Advisory Committee of Ngee Ann Polytechnic's electrical engineering department, Colin is helping guide the future of technological development and the next generation of technologists in the region.

List of Figures

List of Table

Table of Contents

Preface

Revolutions disrupt. They create opportunities and challenges — winners and losers. The fourth industrial revolution is no different, except for the technology involved — we are moving from digital to highly-connected, intelligent, cyber-physical, digital systems. Technology that automates and technology that augments the human-worker in order to elevate efficiency and productivity in all kinds of facilities. Technology that enables new processes and business models that will spur a new age of manufacturing.

As major markets push forward with their Industry 4.0 initiatives, the ASEAN region faces an uphill challenge to stay competitive in this data-driven era. However, playing technological catch up is an important part of achieving the full value potential that Industry 4.0 offers, and as we cover in the following chapters, the ASEAN region has the potential to thrive in this new age.

That ASEAN Industry 4.0 potential is core to our reasons for writing this book, as are the many challenges that lie ahead. Southeast Asian governments and companies cannot rest on their previous successes or be content with regional status. Industry 4.0 disrupts the global manufacturing landscape, automation threatens to shift production facilities back to major consumer markets and those regions that do not evolve will be left behind.

With our combined 60-plus years of experience leading technology development in the ASEAN region, we hope our research and insights will serve to guide readers from their various perspectives, along their Industry 4.0 journeys.

For ASEAN companies, we offer guidance on best practice and common pitfalls when implementing their digital transformation.

For governments, associations, and industry organisations, we analyse and compare the ambitious strategies being put forward, promoting collective ASEAN development.

For international companies and investors exploring opportunities in ASEAN, we provide a full picture of the region, market-by-market, in the context of major global trends.

And for the established and aspiring industrial engineers in the ASEAN region, we present the future.

"Industry 4.0: Navigating the Manufacturing Revolution in ASEAN" takes us through the concept of a revolution to demonstrate the scale of emerging trends. It shines a lens on pioneering European, Chinese, and North American models and considers how we in ASEAN can learn from others to find our place in the global value chain.

We explore the technology behind the smart factory, the critical enablers of Industry 4.0 success, and how manufacturers can identify solutions and strategies that suit their unique needs. We reveal the Industry 4.0 technology trends already taking place across the region and how ASEAN national initiatives are supporting adoption development.

Later chapters offer use cases, case studies, and resource analysis of ASEAN markets. We also present clear guidelines for Industry 4.0 practitioners to consider when looking to increase value from their business in this new world. We hope this depth of research will help all stakeholders navigate change in ASEAN during the fourth industrial revolution.

We invite you to join us for voyage into the future of manufacturing... but first, we should review the past.

For more info visit
www.asean4ir.com

1 Waiting for Industry 4.0

"The Industrial Revolution was another of those extraordinary jumps forward in the story of civilization."
Stephen Gardiner

Walking onto the assembly floor of the VinFast car assembly plant in Hai Phong, Vietnam, and immediately, you notice that this is not a traditional manufacturing operation. The general atmosphere is one of quiet, coordinated motion. Yes, you hear the whine of power tools, but mainly, you notice the orderly stacks of parts and the coordinated movement of humans and machines — 1,200 Swiss-made robots from renowned manufacturer...

You might not notice, but many of the sights and sounds you experience are the results of manufacturing inventions, processes, and ideas developed during the past 270 years. The history of manufacturing technology is a long story, one filled with useful knowledge and social incentives. These drivers led to changes in technology and practices in not one, but three earlier industrial revolutions. Now, companies and governments consider the requirements and potential outcomes of a fourth industrial revolution, also known as Industry 4.0.

Retracing connections between past ideas, incentives, and production capabilities is a useful way to assess business and social challenges we are likely to face in Industry 4.0. So, our modus operandi will be to consider the big ideas, useful knowledge, and incentives of each industrial revolution. Then, we'll see which practices and capabilities emerged from the cauldron of change.

1.1 Industry 1.0: Power & Productivity

When we say, "the Industrial Revolution," what do we mean? There are many definitions, which go something like this one: "The era of economic and social changes, which began with the mechanisation of agriculture and manufacturing, at about 1760 in England and later in other countries."

Given our society's focus on technological progress, it might be helpful to change the emphasis a bit. By moving our attention from the period to the role of change, our description of Industry 1.0 becomes: "The era from 1760 to 1830, typified by steam-powered, factory-based production, in which profound economic and social change became the normal condition."

School Books encourage us to think about the Industrial Revolution, a single-stage event that changed society in one go, but this is not so.

1.1.1 Measuring Time by Technological Change

Historians of technology now recognise not one but three past industrial revolutions. These society-changing surges of technological development (revolutions) are separated by lulls in innovations (evolutions). In evolutionary periods, technological developments are improved little by little and absorbed into standard manufacturing methods.

Where are we today? — Using this model of time and change, modern industrial progress spans 270 years of social, economic, and technological change. Not all historians agree where we are along the industrial revolution timeline. Most consider that in 2020, which we're edging into the fourth industrial revolution (4IR) or Industry 4.0.

Some societies still run their manufacturing businesses with technologies, methods, and ideas typical of Industry 1.0 through 3.0. However, as we'll see in later chapters, playing technological catch up is an important part of achieving the full value potential that Industry 4.0 offers.

Progress as technological leapfrog each surge in manufacturing development is enabled and expanded by innovations of earlier stages. Our walk through the VinFast plant provides glimpses of artefacts from previous revolutions as well as current times: machined car parts (Industry 1.0), assembly lines and electrical power (Industry 2.0), and electronic production controls and robots (Industry 3.0).

1.1.2 Centralising, Mechanising, & Powering Work

In 1760s Great Britain, manufacturing innovators became aware of the business opportunities enabled by two factors of production: a denser-than-water power source and the then-new notion of factory production.

There was no single "aha!" moment in this development. Inventors and investors gradually realised that businesses could fill the growing demand for manufactured goods. To reap the financial benefits, they had to do three things:

- Gather all production resources at one centralised location, the factory, rather than in scattered places at workers' homes.
- Substitute human and water energy with coal-based steam power.
- Speed up the manufacturing process by using special-purpose, machinery rather than human effort.

The result was goods that could be manufactured at an unprecedented speed and scale. For the first time, observers noticed physical improvements in the lives of a single generation of industrial workers

The development of useful knowledge — it's easy to think about the first industrial revolution as being all about smokestack industries invented without the basis of prior knowledge and experience. The truth is, mid-eighteenth-century British society was already technologically competent. Mechanisation was also well underway in Britain's agricultural sector.

As early as the late 1700s, British society had its engineers, mechanics, millwrights, and tinkerers. They all made things —and kept them operating — with their hands, skills and collective experience. They called this "useful knowledge" and made it a pillar of economic and social progress.

This knowledge didn't emphasise scientific theory (although there was always a strong connection between scientists and engineers during this period). Instead, it focused on practice, combining the skills of the artisan with technical knowledge and "the knack" of specific skills needed to design, make and operate machines.

A foundation of technological and financial infrastructure — That close connection between the builders, engineers, and scientists developed within prominent technology hubs. Located at south Lancashire, southwest Scotland, Birmingham, and the West Midlands, these cities and regions became centres of invention and investment. Banks were slow to fund new farm technology ventures. However, merchants, projectors (people who planned and set up projects or enterprises), and speculators picked up the slack.

1.1.3 Early Incentives of Machine Development

The impulse to develop industrial machines grew with the increasing demand for consumer products. The gradual and growing understanding of the role of machines in improving manufacturing productivity was another stimulus.

From the earliest days of the first industrial revolution, money was on everyone's mind. Machine designers, makers, and buyers had two questions: "Would it work?" and "Would it make money?"

Potential buyers had to be persuaded that agricultural machines actually delivered the productivity and profit they promised. A process that began on British farms decades before the industrial revolution. Gradually the successful development of farm equipment and promise of increased cash flow encouraged forward-looking investors to experiment with new machines and processes.

1.1.4 New Energy Sources & Disruptive Ideas

The combination of a new energy source and a new way to organise work provided the foundation of Industry 1.0. Development of coal-powered steam technology and the creation of the factory system are the most familiar parts of the early Industrial Revolution.

Figure 1-1: First Industrial Revolution

The knowledge of builders and mechanics through experience of mechanised agriculture, and the formation of investment and technology hubs are lesser known elements of pre-industrialisation. However, these ideas and events were crucial in helping Industry 1.0 revolutionise manufacturing and laid the foundation for many innovations to come.

1.1.5 Pre-industrial Conditions in ASEAN Nations

Britain began industrialising in earnest in the 1760s, quickly followed by other major European economies of the time. Two centuries later the ASEAN bloc was formed, and it was only then that Southeast Asian countries started their industrialisation process in earnest.

Other than Singapore, manufacturing processes in Southeast Asia at the time were relatively primitive and small-scale. The manufacturing operations were dominated by local natural resource processing. Products included a narrow range of consumer goods sold to domestic markets.

There were a few large modern industrial plants and very little foreign direct investment. Investment activity that existed was a holdover from the colonial era. Manufacturing entrepreneurs in ASEAN nations had to import much of the skills, experience, and funding needed to start local manufacturing sectors.

1.2 Industry 2.0: Power & Process

Developments in Industry 1.0 manufacturing emphasised a stronger power source and a more efficient way to organise manufacturing work.

The next industrial revolution, Industry 2.0, represented a surge of industrialisation from 1870 to 1914. During this time, manufacturing innovation benefitted from the discovery of a newer, more potent energy source and new, more efficient manufacturing and post-production methods.

Also known as the Technological Revolution, Industry 2.0 was a time when the innovations of the past became the useful knowledge of the late 19th and early 20th century. This knowledge, when combined with new ideas and economic and social incentives, enabled manufacturers to produce consumer and capital goods at an unprecedented scale.

1.2.1 Operations take centre stage

In the late 1700s, commercial incentives made manufacturers very interested in the idea of efficiency and how to achieve it. Most of the big ideas driving Industry 2.0 were related to process and ultimately profit.

Electrified manufacturing methods — In the mid-to-late 1700s, coal-based steam power replaced human effort and waterpower during Industry 1.0. Later, energy-dense oil and gas-powered electricity replaced steam power in factories.
Using electricity in manufacturing made business sense. First, it was cheaper than steam. It also enabled a more efficient arrangement of machine tools along the assembly line on the factory floor.

The phenomenon of electricity was old news—incandescent light bulbs hark back to 1809. However, building a useful infrastructure that delivered safe, reliable electrical power was another matter. That accomplishment came in 1882, when figuring out how to deploy and use electricity in factories required decades more time and effort.

Mass production — Steam-powered textile manufacturing made it clear that machines could produce more goods much more quickly than human-based production. Mass production became the umbrella term for the high-volume manufacture of products that used standardised parts, techniques, and machinery.

Experience with the machine tool industry and creating standardised parts (both late-18th-century innovations) provided useful knowledge

7

that enabled the development of assembly-line mass production about a century later.

Assembly line production — Mass production marked the change from time-consuming, file-and-fit production methods to time-saving assembly methods. Assembly line production was another significant change in manufacturing process design. A specialised type of mass production, assembly line manufacturing uses ideas developed before 1870:

- **The division of labour** — the plan that divided a process into different specialised tasks, first described by Adam Smith in his 1776 book, The Wealth of Nations.

- **Standardised, pre-machined parts, machined to specific tolerances** — This approach moved away from the artisan's approach to manufacturing methods. It also minimised the time and effort a worker needed to complete a task. Weapons manufacture in 1790s America was the first application of this production method.

- **Continuous flow production** — The famous Ford Model T moving assembly line used this idea, which already existed in flour mills, breweries, canneries, and industrial bakeries. Ford used the basic assembly line idea developed by Ransom E. Olds, a rival car manufacturer of Oldsmobile fame. Ford improved the process by bringing each semi-assembled car to workers, who repeated the same simple task, time and time again.

The goal of these improvements was to increase production efficiency by minimising the time and effort spent doing each of a series of tasks. Lower costs not only boosted Ford profits but increased revenues, too. Lower cost of production famously made the Model T accessible to more customers and triggered the automobile age.

- **Efficiency** — The idea of efficient operations as a good thing, which manufacturers can achieve with process improvements, also goes back to The Wealth of Nations. However, it took more than a century to take hold in the early 1900s. That's when followers of the Efficiency Movement developed methods that made manufacturing more profitable by avoiding inefficient processes and wasted resources.
- **The supply chain** — This idea rose to prominence the era of assembly-line production. It includes everything—all the people, information, and resources—that manufacturers had to manage in manufacturing. The supply chain was an end-to-end concept, which embraced product sourcing, production, and distribution, from suppliers to customers.

Figure 1-2: Second Industrial Revolution

Development of assembly-line production in the early 20th century put the focus on managing the cost, origin, and quality of parts as well as finished products. Manufacturers started paying attention to supply chains to cut the costs of surplus and shortage of parts.

1.3 Industry 3.0: Process Control & Digitisation

In 1968, development of electronic and IT systems landed the first programmable logic controller (PLC) in Modicon. The advent of the PLC enabled the automation of processes previously controlled by electromechanical devices.

This innovation was a significant indicator of Industry 3.0. Before that, the emphasis on manufacturing was on designing and using increasingly more efficient production processes. Since the beginning of Industry 3.0, however, digital process control has become the critical enabler of the next revolutionary surge of manufacturing innovation and productivity.

Moving from analogue to digital production process control is the hallmark of the third industrial revolution. Electronics and information technology matured as companies introduced semiconductors and later mainframe and personal computers (PCs). Sometime later, automated manufacturing production processes and supply chains went global.

1.3.1 Industry 3.0: Shaping Our World Today

Industry 3.0 manufacturing is part of a familiar world, but its innovations still leave their mark on how we live and work today. For manufacturing, the most influential developments came from digital methods of running, automating, and controlling production processes.

Memory controller unit (MCU) and computers — These now-mature technologies have been around for a half-century. In their current form, they drive the automation of an entire production process and will continue to evolve with the improvement in hardware, software and communication.

Robots — The iconic manufacturing technologies of our time, robots perform programmed sequences of tasks. Completing an entire production process without human intervention is the 'holy grail' of modern manufacturing. While far from an imminent reality, it is one of important goals of Industry 4.0 production innovation.

Digital manufacturing processes and control — Smarter software, more dexterous robots, and new production processes such as additive manufacturing are the pillars of our current era of advanced manufacturing.

When used together, digital processing and control technologies offer the promise of bringing down the costs of smaller, more diverse batches of products. Consequently, this brought about a retail era symbolised by goods customised to buyers' tastes and increasingly personalised demands.

Figure 1-3: Third Industrial Revolution

This more efficient, lower-cost production will not only determine how goods are produced but where. As manufacturing companies use technology to move from mass to customised production, some offshore production is moving back to the countries with higher manufacturing cost.

As labour costs become a smaller part of total production costs, companies can afford to move closer to their customers. Benefits of this new arrangement include lower transport costs and faster response times to changes in demand and customer expectations.

1.4   The Dawn of Industry 4.0

The third Industrial Revolution began in the late 1960s. Experts debate whether we're still in Industry 3.0 or have moved on to an era of more advanced production methods.

Figure 1-4: Fourth Industrial Revolution

Wherever we are in the chronology of industrial development, one thing is certain. Manufacturing process automation enabled by memory-programmable controls and digital technologies has already left their mark on how we produce and distribute goods.

More advanced technologies, which automate an entire production process without human assistance, lie in the province of Industry 4.0. This future development will use production systems controlled by computer technology and humans working together. Furthermore, the volume of autonomous production will expand when processes connect via the internet and internet-ready assets.

These new systems, which support communication between manufacturing processes and facilities, are the next step in production automation. That means the day of smart factories and fully automated, hands-off manufacturing are possible and might be coming soon.

Figure 1-5 shown the development of various industrial revolution in term of production system, technology, competitive priority and manufacturing concept from 18th to 21st century.

	Industry 1.0	Industry 2.0	Industry 3.0	Industry 4.0
Timeline	18th Century (1784)	19th Century (1870)	20th Century (1969)	21st Century (Today)
Production System	First Mechanical Loom	First Production Line Cincinnati Slaughter House Car assembly	First Programmable Logic Controller (PLC) Modicon 084	Cyber Physical System (CPS)
Technology	Introduction of water and steam powered mechanical manufacturing system	Introduction of electrically powered mass production based on the division of labour	Uses electronics, IT and OT to achieve further automation of manufacturing	Convergence IT and OT, autonomous machine based on Cyber-Physical Systems (CPS)
Competitive Priorities Evolution	Quality, Cost	Quality, Cost, Time	Quality, Cost, Time, Flexibility	Quality, Cost, Time, Flexibility, Innovation, Adaptability
Manufacturing Concept	Mass Manufacturing	Moving Assembly Line	Mass Customisation	Mass Personalisation

Figure 1-5: From Industry 1.0 to Industry 4.0

2 International Rush to Industry 4.0

"In the industrial revolution Britain led the world in advances that enabled mass production: trade exchanges, transportation, factory technology and new skills needed for the new industrialised world."
                                                    -Lucy Powell

For many, Industry 4.0 is an inescapable buzzword, a nexus of hype or a business opportunity, while some others see it as an initiative, a technological call to action.

Sceptics may see Industry 4.0 as an umbrella term for a growing number of business and political challenges, while futurists may only see the boundless potential that this new era offers.

2.1 Industry 4.0 Initiatives Backgrounder

Although the content of Industry 4.0 initiatives varies quite a bit, it tends to answer several basic questions, which we present here:

2.1.1 What's the Hurry to Industry 4.0?

Countries throughout the world recognise Industry 4.0 technologies as potentially disruptive forces that can produce business profit or pain. Many nations identify them as time-sensitive drivers of economic growth and technology development. The business opportunities that Industry 4.0 technologies enable are unmistakable. Unfortunately, the window of opportunity that manufacturers can use to compete more effectively is short.

Think of Industry 4.0 adoption as a steeple chase. Manufacturers who are quick to start and nimble over the obstacles will gain a competitive advantage over their rivals. That means assembling the political will, resources, and persistence needed to prevail over legacy technology, practices, and thinking. Finally, national leaders must make these daunting decisions at the high speed of 21st Century technological development.

2.1.2 The Common Thread

What these national frameworks have in common – with each other and ASEAN initiatives – is a focus on investment in people, education, research, and technology. The goal: to develop and take advantage of digitised manufacturing equipment, solutions, and processes to make manufacturing more efficient and workers more productive.

The importance of embedded devices and intelligent non-human process control in national plans is often manifested in high-profile "smart" initiatives. These would include as smart factories or productions, designed to drive operational efficiencies and ultimately stimulate economic growth.

In this chapter, we review the national initiatives of ASEAN member states with major manufacturing operations. We also describe approaches from several established manufacturing heavyweights around the world.

2.1.3 What is an Industry 4.0 Initiative?

National Industry 4.0 initiatives are plans or policy documents that describe each country's vision of the opportunity and challenges that these disruptive technologies present to manufacturers. Governments distribute these plans as one or several documents. Industry 4.0, however, means different things to different stakeholders.

To governments — Industry 4.0 initiatives represent important priorities and development goals to help identify changes in policies, incentives, and ideal standard practices.

To industrial companies — initiatives identify new opportunities brought forth by the real-time access to the intelligence, and point to incentives and sources of investment to transform the way they conduct the business.

To academic institutions — initiatives help identify which technologies or resources need to be strengthened to support innovation during or initial research or advanced R&D phases. Educators can use national development plans to determine which types of training and education to develop or offer to non-university students.

To the individual — Established or newly graduated engineering professionals can use initiatives to identify likely trends in manufacturing tech development, education, and employment. Initiatives are a great place to start looking for indicators of current and future needs for manufacturing professionals and technical specialists.

2.1.4 Why Bother Creating an Industry 4.0 Initiative?

Industry 4.0 plans are public statements, which ideally identify and summarise all the critical points that stakeholders need to implement a program successfully. That means describing and putting stakeholders' seal of approval on:

- **High-level vision that shapes the overall plan** — The nation's overriding outlook about manufacturing technology and its role in transforming the economy, for example.

- **What's most important to those stakeholders** — Initiatives set goals and prioritises the industry's future approaches to achieving them.

- **Methods used to achieve goals** — Clearly defined goals help all stakeholders envision the future but presenting the methodology helps guide them through the entire process.

- **Fair representation** — Any good initiative should assign roles to government, business, and academic stakeholders.

These are fine as general principles go, but which political, economic, and technological trends drive countries to adopt and implement Industry 4.0 tech and ideas?

2.1.5 Drivers of Industry 4.0 Strategy

Embracing Industry 4.0 technology and processes is driven by a multitude of compelling factors. On the more idealistic end of the motivation spectrum, it encompasses the ambition to elevate a nation's manufacturing capabilities, propelling them up the value chain. This ascent is fueled by the adoption of cutting-edge technologies, which, in turn, bolsters productivity and sharpens the competitive edge. Moreover, it serves as a catalyst for the transformation of the workforce, nurturing a more skilled and adaptable labor pool.

This paradigm shift involves trading in the traditional assembly line production for a new role as a provider of automated, highly customized products, and the associated array of tailor-made customer services. The allure lies in the ability to meet individual customer demands with unparalleled precision and efficiency.

Beyond the realm of idealism, the prospect of revitalizing stagnating manufacturing performance beckons as a tangible and pressing goal. Industry 4.0 introduces efficiency-inducing technologies that act as the elixir of rejuvenation for legacy manufacturers, enabling them to reclaim their competitive prowess. This modernization not only ensures survival in a dynamic marketplace but positions these manufacturers as formidable contenders in the ever-evolving landscape of industry.

2.1.6 How Do We Know It works?

You might wonder how to know whether anything meaningful will result from the hype and noise of Industry 4.0 marketing. The answer is simple. We'll know when nations and businesses get tangible indicators of value. That means value measured as higher revenue and lower costs of manufacturing operations; greater worker productivity; and more resilience to changes in the global manufacturing marketplace.

These crucial overriding elements influence the direction of the initiatives but significant insight can also be gained from a deep dive into specific initiatives from ASEAN member nations and beyond.

2.2 National Industry 4.0 Initiatives

Industry 4.0 is the data- and technology-intensive transformation of manufacturing and other related industries. These changes are set in business and technical environments that link data, people, machines, processes, services, and IIoT-connected devices.

The history and current status of Industry 4.0 efforts in countries within and beyond Southeast Asia show their concerns and priorities about current and future manufacturing development. We begin our analysis at the birthplace of "Industrie 4.0", Germany.

2.2.1 Germany: Industrie 4.0 Initiatives

Before there was Industrie 4.0, there was "Industrie führ punct null" or "Industry Leader point zero".

The Industrie 4.0 idea began in 2006, even though the name didn't exist yet, and digitisation was a long, long way from public consciousness. Industrie 4.0 is the name given to a German strategic initiative introduced in 2011. Industrie 4.0 was developed by the BMBF (Ministry of Education and Research) in Germany to create a coherent policy framework that would maintain the nation's industrial competitiveness. The strategy emphasised strong product customisation enabled by intelligent, highly flexible mass production.

Industrie 4.0 embraced these six design principles, which later played essential roles in Industry 4.0 thinking:

- **Interconnection** — Devices, machines, sensors, and humans communicate and connect via the Internet of Things (IoT).
- **Information transparency** — Vast amounts of useful information that help manufacturers make more accurate and timely business decisions.
- **Inter-connectivity** — The ability of operators to connect devices, machines, IT networks, and people is the basis of smart manufacturing. These connections enable the collection and transfer of immense amounts of data from all points in the expanded Industry 4.0 manufacturing process.
- **Technical assistance** — Industrie 4.0 recognises two types of assistance systems. First, assistance systems, which support humans by visualising and gathering information taken throughout the manufacturing process. Manufacturers and partners throughout the value chain use this information to make informed decisions and solve urgent problems, even at short notice. Also, cyber-physical systems conduct a wide range of support tasks, the jobs that are too exhausting, unpleasant, or unsafe for humans to complete themselves.
- **Decentralised decision making** — This occurs when cyber-physical systems make decisions and perform tasks with little or no human attention. Ideally, tasks are only delegated to humans in cases of interferences or conflicting goals.

- **Sustainability** — Industrie 4.0 envisions environmentally sustainable manufacturing, which uses green manufacturing processes, supply chain management methods, and products.

Smart factories, another important Industrie 4.0 idea, are where the six design ideas come into play. It's where data, network connectivity, and cyber-physical systems interact in processes that occur throughout manufacturing supply and value chains.

These ideas and German leadership in the global Industry 4.0 development effort have merged into global Industry 4.0 events and practices. German businesses participate in Open Industry 4.0 Alliance meetings at the Hannover Messe trade fair.

Now more than ever, Industrie 4.0 events and publications describe tech-wonk implementation topics. AI as the driving force of autonomous systems, movement to the mittelstand (SMEs), and intelligent support of end-to-end manufacturing processes are standard bills of fare.

2.2.2 China: The Made-in-China 2025 Plan

Awash in a sea of international high drama, China's Made-in-China 2025 initiative stimulates a range of emotions. At first, Chinese manufacturers and government officials were filled with pride and promoted the plan vigorously. Some of the country's economists tried to inject a more measured response with less ambitious goals. Overseas, industrial rivals, chiefly in the United States, were sure that the Chinese were aiming for world domination in manufacturing. Beijing insists that it is just a guide meant only to steer Chinese development into the future.

The Made in China 2025 plan was drawn up by more than 150 Chinese scientists and scholars in 2014. The Ministry of Industry and Information Technology (MIIT) and 20 other cabinet-level agencies supervised its creation. In 2015, vice-premier Ma Kai led a group that coordinated the planning and implementation of MIC2025 policies nationwide. A technology roadmap published a few months later provided more specific targets for industries mentioned in the plan.

The launch of MIC2025 has been driven by the nation's concern on both its weakness in core manufacturing capabilities and ambition to catch up with the leading players in international arena — Like many manufacturing rivals , China worries about the middle-income trap, although at a different scale and higher up the value chain as the 2nd largest economy of the world. Chinese manufacturers and government officials worry about continued reliance on foreign technology in its supply chains and there's a growing sense of national pride in domestic technology companies.

Largely state-driven, the MIC2025 plan wants to move the country's role away from being the world's assembly line. The plan's goals include moving Chinese industries up the value chain by replacing imported components and technologies with locally manufactured products. The initiative also aims to develop world-class technology champions, who can take on the Western tech giants in cutting-edge technologies.

Clearly defined targets and a long-term outlook — The original idea behind MIC2025 was simply to catch up with other countries. Somehow the plan's objectives became caught up in recent Chinese nationalism. Now, the desire to be a technology superpower is deeply bound up in "Dream of China."

In detail, the plan aims to end China's reliance on foreign technology and raise local high-tech industries up to Western standards. The plan documents the first of three 10-year periods of digital industrialisation:

- Phase 1: 2015—-2025 (becomes a strong manufacturing country)

- Phase 2: 2025—2035 (able to compete with developed manufacturing powers)
- Phase 3: 2035—2045 (transformation into a leading manufacturing power)

In the Chinese government's official view, the plan is a market-led effort, guided by the national government. Details of the initiative present a significant leapfrogging from previous economic models.

The MIC2025 plan provides specific targets for 2025. These include:

- 70 per cent local market share of total domestic technology suppliers.
- Lower manufacturing operations costs
- Shorten development cycles for Chinese manufactured goods
- Product defect rates reduced by 50 per cent
- Establish 40 national (and additional regional) innovation centres.

What's the point of all the time, treasure, and national prestige put behind the China 2025 program?

Targets and priorities — To be sure, the state's involvement in industrial technology development has grown in the past decade. Like earlier Chinese development plans, MIC2025 lays out specific growth and market share targets. In some cases, these goals include 95 per cent domestic market shares for particular industries.

Key Sector	Technology/ Production Target
Advanced information technology (AI, quantum computing)	Home-made chips used in smartphones to make up 40 per cent of the local market by 2025.
	Domestic firms to have 60 per cent of the market in industrial sensors.
Automated machine tools and robotics	Robots should make up half of the domestic market by 2020 and 70 per cent by 2025.
	The country is aiming for two or three local technology champions.
New materials	Advanced basic materials, such as those used in construction or textile manufacture, should hold a 90 per cent share of the domestic market by 2025.
	Essential strategic materials, including rare earth and special alloys should have 85 per cent share of domestic content by 2025.
Biopharma and advanced medical products	China wants home-grown drug firms to be up to international standards by 2025.
	By 2025, 5 to 10 locally developed drugs will have won approval in the United States or Europe.
	Chinese companies will capture 70 per cent of the market for middle- and high-end medical equipment used at county-level hospitals.

Table 2-1: Make In China (MIC) 2025

Here are some sample targets for specific industries chosen for their relevance to manufacturing:

The plan also includes these rather general high-priority tasks to promote breakthroughs in 10 key areas:
- Improve manufacturing innovation
- Integrate technology and industry
- Strengthen the industrial base
- Promote Chinese brands
- Enforce green manufacturing
- Restructure the manufacturing sector

- Promote service-oriented manufacturing and manufacturing-related service industries
- Internationalise manufacturing
- Develop the resources needed for broad Industry 4.0 adoption
- Address gaps in China's scattered value chain

Regarding building the necessary research community, experts suggest a lack of theoretical scientific knowledge, a skills deficit in some areas, and insufficient patience and perseverance to see projects through in others.

"Although China has stressed the importance of fundamental research in becoming a manufacturing superpower... the proportion of around 5 per cent of overall research and development expenditure since Made in China 2025 was introduced is still small. It's only a quarter to a third of those in developed economies such as the US," points out Liu Yadong, editor-in-chief of Science and Technology Daily.

For the compelling reason of developing a vigorous, collaborative R&D environment, China's poor IP protection undermines healthy R&D competition. Successful R&D requires more open collaboration than in the past, and cross-company collaboration is still limited. When collaboration does occur, it is mandated or encouraged by the government. This peer-to-peer distrust creates problems for future development because current industrial innovations tend to cross traditional boundaries of expertise.

It's also worth addressing the need of establishing the right roles between the private and public sectors. China's government has achieved remarkable economic growth with a tightly disciplined management style that's typical of a top-ranked enterprise. To reach the next stage of economic development, however, China's private sector will need to play a much stronger part.

Next, we review the Industry 4.0 initiatives of another nation, known for its powerful and independent private sector—the United States.

2.2.3 United States: Industrial Internet Initiatives

In the coming decades, businesses will establish global networks of machinery, warehouse systems, and production facilities by connecting physical, computer-controlled, and intelligent cyber-systems. And in the manufacturing environment, these cyber-physical systems will include smart machines, storage systems, and production facilities. These networks will exchange information, trigger actions, and control each other with varying levels of independence.

The U.S. name for a group of disruptive and possibly transformative technologies is the Industrial Internet, more recently known as the Industrial Internet of Things (IIoT).

A bit of Industry 4.0 history in the United States - The U.S. Industry 4.0 initiative supports fundamental improvements in manufacturing, engineering, materials purchasing, and supply chain management. The initiative is the product of several programs:

The Advanced Manufacturing Partnership (2011) was a national effort that brought together representatives from industry, universities, and the federal government. Its goal was to identify challenges and opportunities to transform the technologies, products, and processes across several manufacturing industries.

- Launched in April 2014, the Advanced Manufacturing Partnership (AMP) 2.0 is a national effort appointed by President Obama to secure U.S. leadership in emerging technologies. It aims to create high-quality manufacturing jobs and enhance America's global competitiveness.
- Ensuring Leadership in Advanced Manufacturing (2011) called for a partnership between government, businesses, and

educators. The plan identified the most pressing manufacturing challenges and business opportunities.

- Capturing Domestic Competitive Advantage in Advanced Manufacturing (2012) described the policy opportunities and requirements of the United States as viewed from economic and national security perspectives.
- The National Strategic Plan for Advanced Manufacturing (2018) documents opportunities for federal policy to accelerate the development of advanced technologies in manufacturing.

Each of these programs focused on developing stronger ties between expertise, innovation, and a strong economy.

Innovation, talent, and a healthy business climate — Objectives and recommendations related to Industrial Internet proposals have been scattered throughout many documents. However, the central ideas about what needs to be done to transform U.S. manufacturing are collected as recommendations in Capturing Domestic Competitive Advantage in Advanced Manufacturing - a report released through the Advanced Manufacturing Partnership Steering Committee.

When rendered down to the essentials, the recommendations focus on three basic goals: making innovation faster and less expensive, developing a future U.S. manufacturing workforce, and improving the manufacturing business climate.

Accelerating innovation — The first pillar highlights the need to facilitate R&D to take ideas from the drawing board, lab, or testbed, to saleable products and services. These recommendations directly address the hypercompetitive nature of manufacturing. The landscape of advanced manufacturing is evolving at an unprecedented pace, driven by innovation, emerging technologies, and global competition. To position USA at the forefront of this dynamic sector, a holistic approach is required.

Increasing R&D funding in key cross-cutting technologies is crucial to fuel the engine of advanced manufacturing. This begins with the development of a starter list of these technologies, which are deemed vital to the sector's growth. The proposed Advanced Manufacturing Partnership process serves as a valuable tool for evaluating technologies that warrant R&D funding. This approach aligns resources with innovation, making sure that investments are targeted towards the most promising avenues.

Additionally, the creation of a national network of manufacturing innovation institutes marks another significant stride in advancing advanced manufacturing. These institutes, fostered through public-private partnerships, cultivate regional networks of collaborators, strengthening the knowledge-sharing ecosystem and amplifying the impact of research and development initiatives.

To enhance collaboration between industry and academia, it is imperative to explore changes in the treatment of tax-free, bond-funded university facilities. This can be a pivotal driver in fostering stronger relationships between universities and manufacturing companies. Such collaborations not only bridge the gap between theoretical research and practical application but also stimulate the development of skilled talent ready to contribute to the advanced manufacturing sector.

Facilitating an environment for the commercialization of advanced manufacturing technologies is equally vital. Encouraging manufacturers to participate in university innovation networks creates opportunities for securing capital, particularly in the stages of startup to scale-up. This not only empowers the growth of emerging technologies but also nurtures a culture of entrepreneurship within the advanced manufacturing domain.

Lastly, the establishment of a national advanced manufacturing portal is a practical step towards democratizing access to resources and support infrastructure. This searchable database of manufacturing resources caters to the specific needs of small- and medium-sized enterprises (SMEs), ensuring they can tap into a wealth of knowledge, tools, and partnerships, enabling them to thrive in the competitive advanced manufacturing landscape.

Securing the talent pipeline — This pillar focuses on developing current and next-generation members of the manufacturing workforce. Rapid scale-up of new ideas into production requires a well-trained, innovative, and flexible workforce. The training process involves the next-generation workforce and continuous training of current workers. Addressing public misconceptions about manufacturing is a pivotal step in securing the industry's future. An advertising campaign that generates excitement and interest in manufacturing careers can help dispel outdated notions and showcase the sector's innovation and potential. This campaign not only attracts new talent but also fosters a positive perception of manufacturing as a dynamic and rewarding field.

Tapping into the talent pool of returning veterans offers a dual benefit. Veterans often possess the skills and discipline required to fill gaps in the manufacturing talent pipeline. Connecting them with manufacturing employment opportunities not only supports veterans' transition to civilian life but also provides the industry with a pool of capable and dedicated professionals.

Investing in education, particularly through community colleges, is a proven strategy to address the skills gap in manufacturing. Emulating the best practices of leading innovators, increasing investment in this sector ensures that students receive high-quality training and education, equipping them with the skills necessary to thrive in advanced manufacturing careers.

Providing skills certifications and accreditation is another crucial facet of workforce development. Establishing partnerships to create stackable credentials allows individuals to build a portfolio of skills, making them more versatile and appealing candidates in the job market. These credentials act as a testament to an individual's competency and commitment to their craft.

To advance the education and training of future manufacturing professionals, universities must focus on advanced manufacturing. Developing educational modules and courses tailored to the evolving needs of the industry ensures that graduates are well-prepared to tackle the challenges of modern manufacturing. These programs should be designed in close collaboration with industry partners to ensure relevance and alignment with industry demands.

Lastly, launching national manufacturing fellowships and internships is a practical approach to both improve workforce skills and recognize the career opportunities within manufacturing. These initiatives not only provide valuable hands-on experience but also inspire a new generation of talent to consider manufacturing as a viable and fulfilling career path. By nurturing these programs, we can bridge the gap between education and industry, fostering a symbiotic relationship that benefits all stakeholders.

Improving the business climate — This pillar addresses manufacturing-related tax reform streamlined regulatory policies and improved trade policies. To foster a thriving environment for advanced manufacturing, a multifaceted approach is essential.

Streamlining regulatory policy is a pivotal step in eliminating bureaucratic barriers and facilitating growth. A framework of smarter regulations, tailored to the needs of advanced manufacturing, ensures that businesses can operate efficiently without unnecessary red tape, promoting innovation and competitiveness.

Improving trade policy is equally crucial. By reevaluating and amending trade policies that may adversely affect advanced manufacturing firms, we can safeguard their interests, promote fair competition, and bolster economic growth.

Additionally, updating energy policy is essential for reducing operational costs and enhancing competitiveness. Tailoring energy policies to be more favorable to manufacturing companies ensures a sustainable and cost-effective energy supply, driving productivity and sustainability.

This detailed list identifies the resources and priorities that U.S. stakeholders need to move forward with Industrial Internet proposals. So, how are government, business, and educational stakeholders doing?

Likely challenges to digital transformation — In the United States, the risk profile of IIoT adoption are somewhat different than those of ASEAN member countries. Government-supported research and inter-business collaboration are established parts of technology development. However, there are still resource gaps and obstacles in the United States that will slow the digital transformation process.

General readiness — The biggest challenge of American manufacturers adopting Industry 4.0 technology and processes. Most businesses are not ready. In an Accenture survey of more than 1,400 business leaders, only 36 per cent of respondents claim they fully grasp the implications of the IIoT. While, only seven per cent of companies have developed comprehensive strategies and investments.

A limited view of benefits — As in other countries in and beyond ASEAN, U.S. manufacturers recognise the short-term benefits of Industry 4.0 — more efficient production and more productive employees. The value of Industry 4.0 as an enabler of new and lucrative business models, however, seldom comes up on executives' radar.

Lack of interoperability among existing systems — Integrating manufacturing systems with the internet adds a layer of complexity, which is likely to increase the time and costs of Industrial Internet deployments.

31

Business risks — Another notable barrier is the uncertain ROI on immature or untested technologies.

Finally, we return to Asia review Japan's take on digital transformation. Its broader scope makes it unique among major manufacturing nations.

2.2.4 Japan: Society 5.0

Japan's approach to Industry 4.0 adoption reflects the nation's unique set of challenges and concerns about digital technology changing its businesses and society.

A broader scope is the hallmark of Japan's digital technology adoption plan. Other nations limit the focus of their Industry 4.0 strategy to the digital transformation of manufacturing. Faced with worrisome social challenges, however, Japan developed its Industry 4.0 adoption plan, which goes far beyond the digitisation of the economy. Rather than focusing on changes in technology (industrial revolutions), Japan expands its attention to the effects of digitalised technology on society.

Society 5.0: As the name implies, Industry 4.0 technologies and processes are mainly concerned with industrial production. The fuel in the Industry 4.0 engine is information—data exchanged between people, devices, machines and systems.

The Society 5.0 plan recognises that the availability of vast quantities of data and increasingly powerful analytics are leading to new ways of doing business. However, the plan's authors recognise that changes in technology will also have an impact on social interactions. That's why the plan addresses change in Japanese society.

In April 2016, the Japanese government enacted Society 5.0, the 5th Science and Technology Basic Plan. The plans sets out that today, human still create knowledge from information. In the proposed fifth stage of human society, intelligent machines guided by AI will do the same.

Society 5.0 drivers and objectives — Japanese officials are faced with an aging population and sluggish manufacturing growth. They recognise that digitised processes provide the opportunity to strengthen the national economy, reduce environmental problems, and address social issues.

The Society 5.0 Plan describes high-level capabilities that will enable Japanese citizens to stimulate the development of future industries and reinforce the "fundamentals" of the nation's science, technology, and innovation.

New manufacturing capabilities — The Society 5.0 plan concerns itself with many areas but we'll stay focused on manufacturing. Big data, the Industrial Internet of Things (IIoT), artificial intelligence (AI), and robotics will be incorporated at different levels of manufacturing facilities, machines, and production processes. Within the Society 5.0 framework, digitalised technologies will enable manufacturers to:

- Perform flexible production planning and inventory management in response to changing requirements.
- Make production more efficient by using AI and robots to enable inter-plant coordination and achieve high-mix, low-volume production.
- Make distribution more efficient by using processes such as cross-industry cooperative shipping and truck platooning.
- Provide consumers with highly customised goods with minimal delays in delivery.

Considering the capabilities, we should expect obstacles to digital manufacturing development.

Obstacles to adoption of digitalised technologies — Japan's manufacturing prowess is renowned worldwide, underpinned by its exceptional infrastructure and highly skilled workforce. While the quality of its technically trained employees and engineers is consistently excellent, the full realization of Industry 4.0 technologies demands a culture of innovation. This, in turn, necessitates a technically trained workforce with not only exceptional technical skills but also critical thinking abilities, areas where Japanese workers have room for growth.

Additionally, the evolving landscape of Industry 4.0 places a growing demand on Japanese business owners to possess entrepreneurial skills and instincts, fostering agility and adaptability. The challenges facing Japan in fully embracing Industry 4.0 revolve around refining the infrastructure for technical and business innovation. These challenges are currently the focal point of Japan's efforts to facilitate the seamless adoption of Industry 4.0 technologies. To secure its place as a leader in this transformative era, Japan must continue to cultivate a culture of innovation and entrepreneurship while nurturing the skills of its highly capable workforce, ensuring they can harness the full potential of digitalized technologies.

3 The Industry 4.0 Framework: Idea, Tool, & Guide

"Technology is the future, I have seen the third industrial revolution, and we are in the midst of the fourth industrial revolution."

N. Chandrababu Naidu

Industry 4.0 is many things to many people. For our purposes, Industry 4.0 is a journey deeply involving various advanced technologies that help manufacturing operations become more reliable, productive, efficient, and customer-centric.

Another useful definition of Industry 4.0 (among a multitude of others) is the information-intensive transformation of manufacturing and other industries. The Industry 4.0 environment digitally connects data, people, processes, services, systems, and IoT-enabled industrial assets across cyber and physical worlds. The goal is to create, use, and take full advantage of actionable information.

For some analysts, Industry 4.0 describes a future state of industry characterised by thorough digitalised production processes. For others, Industry 4.0 is already here, representing a new and higher level of organisation and control over manufacturing along entire value chains and product life cycles.

This chapter views Industry 4.0 as the development of a roadmap to establishing high-tech, digital manufacturing processes. We'll give special attention to the framework's reference architecture, which bridges physical industrial assets and digital technologies in cyber-physical systems.

3.1 Pillars of the Industry 4.0 Framework

There are many Industry 4.0 frameworks. Each country engaged in systematically modernising its manufacturing base has its own. As in

Japan (Society 5.0), the scope of the framework might expand beyond manufacturing. National development priorities might focus on different sets of advanced technologies. However, countries engaged in Industry 4.0 programs and initiatives tend to emphasise a standard model, a set of advanced technologies, and concepts.

3.1.1 Technology Pillars of Industry 4.0

Industry 4.0 depends on not one but several advanced technologies. Some are familiar; others have been a commercial product for a short time. It's the combination of these technologies in R&D, production, and post-production processes that will help make manufacturing more efficient.

Different analysts use slightly different lists of technologies. (Ours comes from a 2017 Boston Consulting Group study.) However, these are the technologies usually mentioned in Industry 4.0 frameworks:

- **Big data/advanced analytics** — The industrial world is filled with mountains of unanalysed product and process data. Analysing it and turning it into actionable information can optimise production quality, improve services, and enable faster and more accurate decision making.
- **Advanced robotics** — As robots become more flexible, cooperative, and autonomous, they will interact with one another, work safely with humans, and eventually learn from humans, too. Industry 4.0 provides a manufacturing context for these opportunities.
- **Advanced simulations** — In Industry 4.0 environments, 3D simulation of product development, material development, and production processes will enable operators to test and optimise processes for products before production starts.
- **AI/cognitive computing** — Cognitive manufacturing uses the assets and capabilities of the IoT, advanced data analytics, and

cognitive technologies such as AI and machine learning. When used together these technologies will drive improvements in the quality, efficiency, and reliability of manufacturing processes.

- **Industrial Internet of Things** — In the IIoT, an ever-greater number of products will incorporate internet-connected devices, which link with each other with standard protocols. This approach to manufacturing will decentralise analytics and decision-making and enable real-time responses.
- **Cybersecurity** — Industry 4.0 environments include connectivity and communications protocols as well as sophisticated identity and access management systems. These technologies enable manufacturers to provide secure, reliable communications and data flow throughout Industry 4.0 systems.
- **Additive manufacturing** — In Industry 4.0 manufacturing environments, these technologies are the best choice for producing small-batch, customised, and high-performance products.
- **Cloud-based service-enabling technologies** — Industry 4.0 manufacturing operations require more data sharing across sites and companies than earlier processes do. Shifting data storage and management to the cloud will drive the development of more manufacturing execution systems (MESs) that use cloud-based machine data.
- **Augmented reality** -- AR provides an effective way to represent production processes by overlaying real-world views of production with virtual information. In ASEAN countries, the most likely role of AR lies in training future workers and technicians how production systems behave in real-time.

Figure 3-1 demonstrates the breadth of applications that make up Industry 4.0. In the world of Industry 4.0, technology doesn't operate in isolated factories or assembly lines. In fully realised Industry 4.0 environments, technologies connect with other entities, up and down production hierarchies, along value chains, and throughout product life cycles.

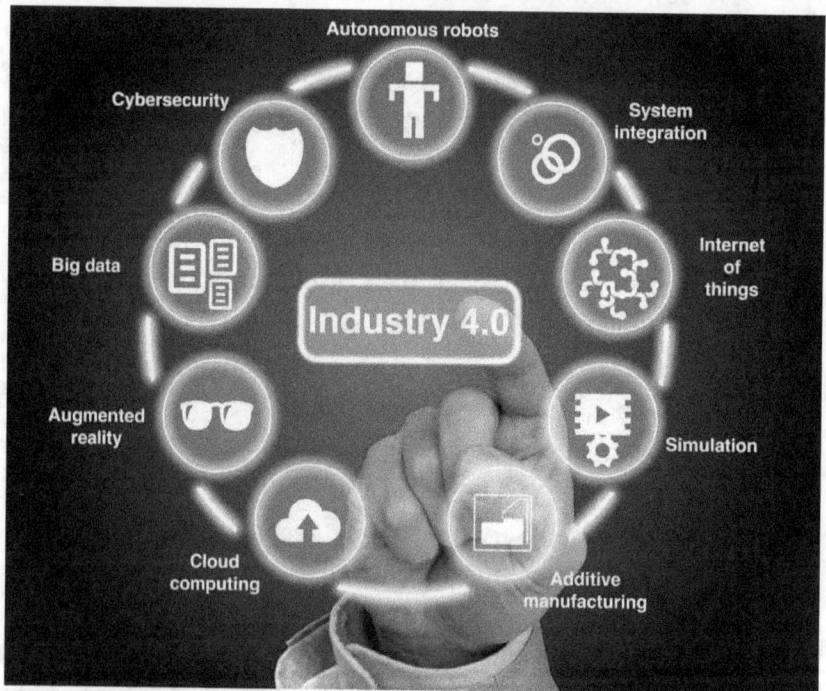

Figure 3-1: Emerging Technologies

3.2 Connectivity

In the globally interconnected world, data sent along digital networks link machines, production objects, internet-connected devices, their virtual representations, and humans. Critically, interconnected machines in Industry 4.0 systems interact with different levels of human involvement. For Industry 4.0 manufacturing and systems engineers, this ever-present connectivity has design and operational implications:

- **Connectivity is related to interoperability** — Shared communication protocols are not just becoming the norm. They are becoming essential parts of manufacturing process design.
- **Connectivity enables cyber-physical systems** — These are the systems that make smart factories possible. Cyber-physical systems connect intelligent production objects to embedded physical devices, which can store and process data.
- **Humans are not always in the production control loop** — Industry 4.0 production machinery no longer simply "makes" the product. The product communicates with the machinery to tell it exactly what to do.

Data that flows through Industry 4.0 systems does so in a systematic way, through production hierarchies, and along product life cycles.

3.2.1 Data integration: the broader I4.0 view

Integration addresses the flow of data between connected machines and devices at different parts of the product life cycle and levels of the production hierarchy.

Horizontal integration refers to the connection of and data flow through IT systems across all manufacturing-related production and business planning processes. Horizontal integration is, therefore, about digitising entire value and supply chains. From supplier to consumer, end-to-end horizontal integration maps IT systems and information flows with big data, analytics, and IoT devices.

In traditional thinking about manufacturing, the production process included all the steps that occur after components enter the factory floor and before they leave it as a finished product. Industry 4.0 concepts require a wider perspective.

Now, a product's life cycle begins with the first product development ideas and extends horizontally through development and production steps to sales and eventual product recycling or disposal.

Vertical integration refers to IT systems connected to machines and devices that operate at different levels of the production hierarchy. In traditional terminology, these hierarchical levels include:

- **Field level** — in which sensors convert environmental data to signals that are analysed and to actuators, which convert signals into actions.
- **Control level** — in which controllers gather process data from sensors and drive actuators.
- **Production process level** — in which automatic devices monitor, control, and adjust specific functions in production processes.
- **Operations level** — which includes functions such as production planning and quality management.
- **Enterprise planning level** — which manages the whole production system, enabling business functions such as production planning and market analysis.
- **Connected world level** — where traditional hierarchy is expanded by moving beyond isolated manufacturing facilities. In this level, network assets and processes connect and support data flow throughout manufacturing systems. Industrial communications networks tie all vertically integrated levels together, sending data from one level of the hierarchy to the other.

The production hierarchy, manufacturing processes, and product life cycle are familiar concepts. In the early days of Industry 4.0, the difficulty lay in how to combine these concepts in a way that was easy to understand and use. The RAMI 4.0 data model helped overcome this problem.

3.3   The RAMI 4.0 Model

The Reference Architectural Model Industrie 4.0 (RAMI 4.0) was developed in Germany as part of the country's Plattform Industrie 4.0 initiative. As Industry 4.0 achieved more acceptances throughout Europe and beyond, the need for a clear and consistent vocabulary became increasingly important. The RAMI 4.0 goal was to create a uniform framework for national and international communications and ideas.

3.3.1   RAMI 4.0 Structure

Targeting at a collective understanding of Industry 4.0, RAMI 4.0 is an idea map that describes manufacturing processes and production objects in a clear and systematic way. RAMI 4.0 ensure that those involved in discussions of Industry 4.0 will understand one another.

In this model, each production object is defined with its related functions and data. The result is a complete, virtual description of the object. The RAMI 4.0 structure is built on a three-dimensional framework consisting of the Value stream and life Cycle, Hierarchy Levels and Layers.

41

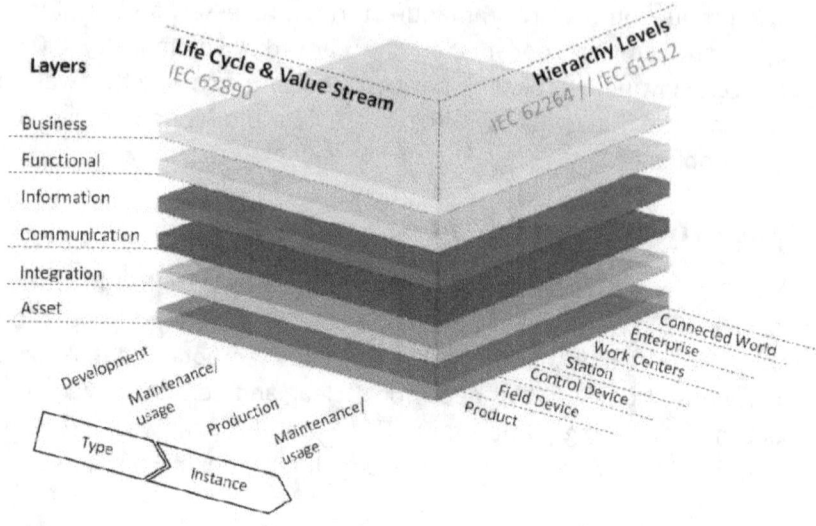

Figure 3-2: Three-dimensional RAMI 4.0 framework
Source: https://www.zvei.org/en/subjects/industrie-4-0

Dimension 1: Life cycle and value stream. In RAMI 4.0, each product is described and tracked from the first idea to the scrap yard by Life cycle axis based on IEC62890 standard. First of all, a specific **product type** consists of an identifier, meta data and associated certificates, while a **product instance** behaves as the instantiation of a product type, characterised by an instance identifier. Based on this definition, the life cycle of a product will detail the life cycles of both product type and instance. According to Rami 4.0 model, product type life cycle starts from the development phase throughout maintenance & usage stage. On the other hand, the product instance life time model starts from the production phase throughout usage stage which includes commissioning and disassembling or disposal of the instance.

Dimension 2: Hierarchy levels. The foundation of the axis description is IEC62264 and IEC61512 (also well known as ISA95 and ISA-88), representing different functional levels of a factory. To make it easier to talk about complex production processes, engineers and plant managers divide them into several categories:

42

- The connected world (new to RAMI 4.0) – visualises and describes the relationship of inter-connected assets, both internally and externally.
- The enterprise – the meaning in Industry 4.0 goes beyond traditional territory of enterprises, referring to both physical organisations and strategic initiatives or missions.
- Work centres – highest level of unified manufacturing production line. A good example can be the stamping line for a typical automotive factory.
- Machines or workstations – refers to the work cells carrying the operations with the resources such as machines, human labours and materials.
- Control devices – characterized by the typical control systems such as PLC and DCS.
- Field devices – field level installation such as sensors and actuators
- Products with expanded scope

Dimension 3: RAMI interoperability layers. This dimension represents different types of data and functions relevant to elements of Dimensions 1 and 2. These data and functions include:

- **Business layer** — represents business-related data exchanged in industrial processes. Allowing users to map regulatory and market-related policies, business models, products, and services of market participants. Data in this layer can also represent business capabilities and processes.
- **Functional layer** — supports the business layer by providing the runtime and modelling environment.
- **Information layer** — describes the data used and exchanged between functions, services, and components. This layer contains the data services such as provisioning and integration. The key value for this layer is its capability of receiving the events from physical asset via lower level layers

43

and applies the adequate processing and transformation to support the upper levels.

- **Communications layer** — emphasises protocols and mechanisms for the interoperable exchange of information between components. The outcome is the unified data formats and interfaces that grant the data access, which has been the bottom neck to the Industry 4.0 adoption for long.
- **Integration layer** — describes physical assets as their digital equivalents. This layer shoulders the most important responsibility of representing the transition from physical world to the cyber space via various innovative approaches (comparing with traditional integration methodologies) to work on documentation, software, control and monitoring mechanism.
- **Assets layer** — identifies and describes the real assets in the physical world.

With precisely defined contexts for Industry 4.0 ideas and production objects, users can work their way through the model knowing that other users have the same information, vocabulary, and contexts.

3.3.2 The RAMI 4.0 Standards

Developing consistency across RAMI users is essential to gain the most from the framework. For this, standards must be developed, adopted, and integrated into systems. In RAMI 4.0 context, there are a few standards that have been leveraged, expanded and presenting integrated value for Industry 4.0 development.

IEC 62264

IEC 62264 showcased in RAMI 4.0 architecture model is a standard for enterprise-control system integration, built on the well-known ANSI/ISA-95. Referenced in Figure: 3-2. ISO/IEC 62264 is an integral

44

part of the RAMI model for Industry 4.0 development and has been singled out as a key standard for the factory of the future initiatives.

The ANSI/ISA-95 (better known as ISA-95) is an international standard developed by the International Society of Automation (ISA). ISA-95 provides standards for the development of automated interfaces between enterprise business and manufacturing control systems.

The ISA-95 framework is used by manufacturers around the world to develop consistent data models and terminology. The primary goal of the standard is to enable smooth information flow across Enterprise Resource Planning (ERP), Manufacturing Execution System (MES) and Supervisory Control And Data Acquisition (SCADA) systems. ISA-95 supports interoperability in all industries and to every type of manufacturing process.

ISA-95 provides the concept of modular segments to define any manufacturing task. Where process designers can link several segments into an operation and perform, track, and schedule more complicated tasks. New generations of IIoT solutions and devices are coming to market and playing a larger role in factory operations. Device evolution is expected to flatten the structure of the ISA-95 model further.

IEC 61512:

IEC 61512, commonly referred to as ISA-88, plays a pivotal role in the realm of batch process control. This international standard, formally known as IEC 61512-1, provides a structured framework for the description and management of equipment and procedures within batch processes. It's essentially a roadmap for optimizing and standardizing the way industries handle batch operations.

At its core, ISA-88 is designed to improve the efficiency, flexibility, and quality of batch manufacturing. It achieves this by establishing a standardized language and set of guidelines that enable industries to design, implement, and operate batch processes consistently. By defining a common terminology and approach, it minimizes miscommunication and misunderstandings, which are often at the root of inefficiencies and errors in batch operations.

ISA-88 addresses both the hardware (equipment) and the software (procedures) aspects of batch processes. This comprehensive approach ensures that all elements of a batch system, from the physical equipment to the control logic and sequencing, are well-defined and well-coordinated. This standard also provides guidelines for defining batch recipes, which specify the precise steps and conditions necessary to produce a batch of a particular product.

By adhering to ISA-88, industries can benefit in several ways. It enables them to be more agile in responding to market demands, reduce batch production times, lower production costs, and improve product quality and consistency. Additionally, it facilitates regulatory compliance by providing a structured framework for documentation and traceability, which is especially crucial in industries such as pharmaceuticals and chemicals.

IEC 62890:

IEC 62890 is an essential international standard that plays a crucial role in the life-cycle management of systems and products used in industrial process measurement, control, and automation. This standard, which is formally titled "Life-cycle management for systems and products used in industrial-process measurement, control, and automation," provides a comprehensive framework for the entire life cycle of these systems and products.

The reference structure outlined in IEC 62890 encompasses various phases and processes, each of which contributes to the efficient and effective management of industrial process measurement and control systems. Here's an elaboration of its key components:

- Concept and Feasibility: This phase involves the initial conceptualization of the system or product. It includes defining the project scope, feasibility studies, and identifying the functional and technical requirements. This early planning is critical for setting the project's direction.

- Specification and Design: In this phase, detailed specifications and design plans are created. This includes defining the system's architecture, hardware, software, and interfaces. The goal is to ensure that the system meets the identified requirements.

- Implementation and Testing: Here, the system or product is built and thoroughly tested to ensure that it functions as intended. This phase involves hardware and software development, integration, and rigorous testing to identify and rectify any defects or issues.

- Installation and Commissioning: Once the system or product is deemed ready, it is installed at the intended location. Commissioning involves ensuring that the system works correctly in the real-world environment, including calibrations and performance assessments.

- Operation and Maintenance: This phase covers the ongoing operation of the system or product. It includes routine maintenance, monitoring, and troubleshooting to ensure continuous performance. Regular updates and adaptations may also be necessary to keep the system current.

- Modernization and Upgrade: As technology evolves, systems and products may need to be modernized or upgraded to remain relevant and efficient. This phase involves assessing and implementing necessary changes to meet new requirements.

- Decommissioning and Disposal: At the end of the system's life cycle, decommissioning and proper disposal are essential. This phase ensures the safe removal and disposal of hardware, software, and materials, with consideration for environmental and safety standards.

- Documentation and Records: Throughout the entire life cycle, comprehensive documentation and records are maintained. This includes design specifications, testing reports, maintenance logs, and any changes made to the system or product.

IEC 62890 emphasizes the importance of effective communication and collaboration among various stakeholders involved in the life-cycle management process, including designers, engineers, operators, and maintenance personnel. It promotes the use of standardized processes and documentation to enhance the reliability, safety, and sustainability of industrial process measurement, control, and automation systems and products.

4 Industry 4.0 & the Global Value Chain

"The digital supply chain has the potential to completely change supply chain. However, do your due diligence before rushing in."

EverythingSupplyChain.com.

In the value chain, every step of each process centre on a single goal: Increase value to the customer or improve the manufacturer's advantage in the marketplace.

The vision of Industry 4.0 highlights technologies that enable greater competitive advantage. Smart factory design and function enable significantly higher productivity, efficiency, and self-managing production processes. In these production environments, people, machines, equipment, logistics systems, and work-in-process components communicate and cooperate with each other directly.

In Industry 4.0 manufacturing value chains, product development, production, and logistics processes combine intelligently across company boundaries. In Industry 4.0 manufacturing environments, forward-looking enterprises don't fixate on speeds and feeds of the machinery. Instead, they focus on how smart assets and manufacturing methods can drive revenue.

4.1 Digitisation versus Digitalisation

Many manufacturers engaged in pioneering Industry 4.0 adoption efforts have arrived at an important discovery — the real value of Industry 4.0 manufacturing lies in moving beyond digitisation to digitalisation. The distinction between these terms is not always clear, even Industry 4.0 practitioners confuse them.

Digitization: Digitization is the process of converting analog or physical entities, such as microfilm images, paper documents, photographs, and sounds, into a digital format, represented as bits and bytes. In smart factories, digitization involves creating digital replicas of physical components, products, or processes. This transformation enables easier storage, retrieval, and manipulation of data, enhancing accessibility and efficiency.

Digitalization: Digitalization, on the other hand, takes the process a significant step further. It involves the integration of digital technologies into various aspects of business operations, functions, models, or processes. While physical entities can be digitized and represented by digital twins, the core focus of digitalization is on optimizing and improving business processes through the strategic use of digital technologies. This optimization leads to process-level improvements that yield maximum value at the lowest possible total cost, creating a competitive advantage for the organization.

Digitalization is not limited to replicating physical objects but aims to leverage digital technologies for enhancing and transforming entire business activities. It is the pathway through which organizations evolve from incremental process-level improvements to achieving broader digital business initiatives and embarking on a journey of digital transformation. The ultimate goal of digitalization is to drive efficiency, agility, and innovation throughout an organization, positioning it for sustainable growth in an increasingly digital world.

In the world of digital manufacturing, transformation is the Promised Land, where good things start happening. Manufacturers discover new business opportunities and create new revenue streams and offerings.

However, these new symbols of a healthy digitalised manufacturing environment come with a hefty price tag. Manufacturing processes will have to be changed. Risk adversity will have to be overcome. And existing partner relationships will have to be expanded or changed.

In short, the price of successful digital transformation is the willingness to change.

4.2   Value Proposition of Industry 4.0

Industry 4.0 has changed how we view business value and the potential to increase it. A growing number of economic analysts suggest that manufacturers have already reached the end of the line for value from traditional cost-cutting measures.

4.2.1   Looking for Value in All the Right Places

A recent online publication added a seldom-mentioned driver to the list of urgent reasons to modernise manufacturing. It appears that concentrating on low wages and cutting production costs have taken manufacturers as far as they can go. Now, they must find new ways to generate value.

Manufacturers throughout the world are eager to benefit from the business value that Industry 4.0 might generate. ASEAN manufacturers hope that Industry 4.0 technologies will help them stay out of the middle-income trap and provide healthy profits for the foreseeable future. It's hard to ignore the McKinsey & Company's eye-popping US$216 billion to US$627 billion ASEAN market forecast that's all over the online press.

Successful adoption of Industry 4.0 technologies will involve looking for value in some new places, such as in customised products, services, and product design. Making it logical to consider value in the global manufacturing environment and the types of benefits that manufacturers might expect from participating in digital transformation efforts.

4.2.2   The Smile Curve & the Outcome Economy

Careful study of the global value chain and classic smile curve as it applies to ASEAN nations provides a vibrant story of international competition and economic development. Advanced countries such as the United States, Japan, and the major EU economies, operate at the high-value end of the product life cycle. These vary from initial concept, design, and branding on one end to marketing, sales, and post-sales services on the other.

Moving up the sides of the smile curve — During the past 30 to 50 years, countries with high-value economies outsource the lower-value work in the middle of the smile—manufacturing—to less developed countries such as ASEAN-6 nations and more recently China.
Faced with recent increases in labour costs, China is joining the move up the value chain. Now, they too are outsourcing lower-value manufacturing work to ASEAN nations with growing workforces of young people and lower operating costs.

The big "if" is digital manufacturing technologies and methods. The regional Industry 4.0 environment enables ASEAN manufacturers to accept higher-value work. However, the seemingly golden opportunity carries a very big "if." To remain competitive, ASEAN-6 (Indonesia, Malaysia, Philippines, Singapore, Thailand and Vietnam) manufacturers must train or retrain their workforce to competently operate modern manufacturing equipment and processes, and they must complete the training quickly.

The outcome economy — No matter how much value Industry 4.0 modernisation might enable, the global manufacturing environment is changing hard and fast. For ASEAN member countries and many other nations, the competitive environment is becoming harsher and less predictable than ever. Even primary considerations such as what to manufacture and how to satisfy customers are changing.

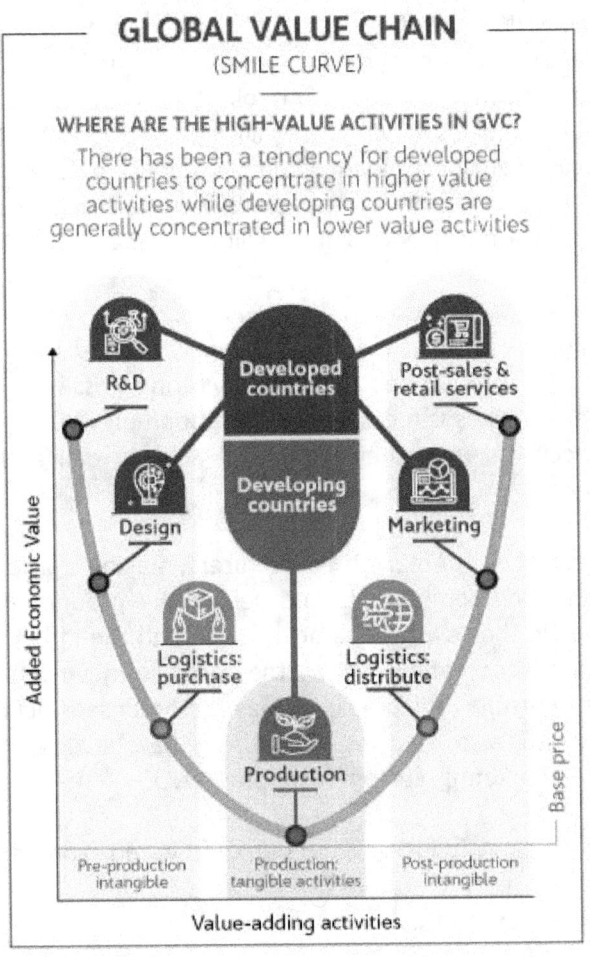

Figure 4-1: The Smile Curve and Global Value Chain

In a May 2015, CIO Journal article, Accenture CEO Paul Dougherty welcomed readers to the outcome economy. This idea described a new business environment. In it, companies create value not by making and selling products but by delivering solutions that "directly produce quantifiable results (outcomes)."

This idea is not new. It's part of the move toward customer-centred marketing and selling that's been in progress for about five years. What makes the outcome economy notable are its enablers — modern digital technologies. Five years ago, easy access to the Internet helped sales prospects learn about a product before they ever met a sales rep while sales professionals were learning to pay attention to their customers and deliver solutions that customers want.

Now, digital technologies such as the IIoT, data analytics, and machine learning are making customer-centred manufacturing possible. The challenge for manufacturers is focusing their attention on which their customers are, deciding what they want and how to deliver the products they want.

Hitting a moving target — For many manufacturers, making a product still involves assembling components and selling the product. The product's value lies in the selling price of the item and the efficiency of the manufacturer's operations. Now, the focus is moving from making a product to results (outcomes) such as customer satisfaction, safety, or comfort. It also involves monetising new stages of a product's life cycle (marketing, customer support, etc.).

For manufacturers in many industries, the outcome-based economy is about:
- Bundling products and services.
- Moving from one-time to ongoing transactions.
- Enabling subscription-based transactions based on a recurring revenue relationship.

In short, the outcome economy emphasises monetising the entire product life cycle, that is, everything about the product.

4.2.3   Finding the Value in Manufacturing

It's well enough to say a manufacturer must start selling more than their finished goods. Unfortunately, it's more complicated.

Business leaders are learning where the value in manufacturing lies and it's not where they might have expected. The lowest part of added value lies in the manufacturing process itself. Much of the value in a manufactured product lies with concept, R&D, branding and product design at one end and distribution, marketing, and sales and services at the other. The smiling curve (Figure 4-1), created by Acer Inc. founder, Stan Shih, in the early 1990s, illustrates this principle.

4.3   Three Levels of Value

When people talk about Industry 4.0, they often mention a two-tier value regime. The first category is what people typically think about value — cost savings, improved worker productivity, and faster times to market. These factors all depend on producing more products, more efficiently. Some analysts call this low-hanging fruit, which can be achieved by improving current manufacturing processes to make products with traditional business models.

Value in the second tier, however, depends on new business opportunities. This is where new and existing Industry 4.0 technologies combine to create new business models and revenue. Finding value in this new frontier of manufacturing requires modern equipment, high-risk tolerance, and a sharp eye for business opportunities

Industry 4.0 technologies help manufacturers with the right stuff to create value at the production process, business, and market levels.

Manufacturing-related, process-level value — This level describes value generated by process operations and maintenance within businesses:

- Manufacture, maintenance, supply, and distribution costs reduced or avoided.
- Improved supply chain transparency
- Lower inventory costs
- Avoiding equipment repair and maintenance costs.

Business-level value — Our next focus of interest lies several levels up the manufacturing hierarchy from processes to the enterprise (at the RAMI 4.0 business layer). This is where you find the total value of products and services generated by a business. The examples are familiar to every MBA student:

- Revenue added via new products, services, or business opportunities
- Revenue recovered from productivity lost to inefficient operations
- Higher revenue, enabled by improved productivity
- Improved production and worker productivity
- Improvements in the quality of goods
- Greater customer satisfaction and loyalty
- Faster product time to market or time to service
- Faster response to customer preferences and requirements

These types of value aren't just measurements of financial performance. They also serve as indicators of potential value growth and business opportunity.

National-level value — Just as value, in general, can indicate opportunities at the business level, value in national level manufacturing activities identifies benefits beyond the enterprise. These difficult-to-quantify metrics reflect improvements in education, technological development, and quality of life:

- Provide modern consumers with satisfaction enabled by advanced products and services that they crave.
- Improve wages and living standards for the nation's people.
- Develop a workforce with skills in new fields such as data science and analytics, data engineering, and agile manufacturing applications.
- Improve innovation in fields such as analytics and advanced production automation
- Create a strong local demand for Industry 4.0 components (IoT, analytics, etc.), which helps companies build capabilities in these areas.

Moving on from business theory, in the next chapter we review the real-life efforts of ASEAN manufacturers and governments to adopt and use Industry 4.0 technologies.

5 The Smart Factory

"We have to prove that digital manufacturing is inclusive. Then, the true narrative will emerge: Welcome, robots. You'll help us. But humans are still our future."

Joe Kaeser

Today, technology is quickly changing manufacturing from an industry of mass-produced goods to customised products. The ultimate business goal—and pillar of 21st century competitiveness—is for manufacturers are to make the right product, deliver it to the right customer for the right price. Ideally, the product will also offer greater utility or customer appeal at a higher level of sophistication.

The Industry 4.0 vision adds intelligent manufacturing systems to the fundamental processes of fabrication and assembly. In this vision, "Digital Transformation" represents a fully connected manufacturing environment. All equipment in this environment is online, intelligent, and capable of making decisions with varying degrees of autonomy.

Earlier chapters of this book present ideas, tools, and approaches to Industry 4.0 that ASEAN member countries and other nations are beginning to take. What's the point? Where do the resources, tools, and ideas come together? And for what purpose? This chapter highlights one of the critical pillars of Industry 4.0 - the smart factory.

5.1 The Smart Factory in Digital Manufacturing

The smart factory idea is central to Industry 4.0., many people even use the terms interchangeably. Unfortunately, there's little consensus among analysts and manufacturers as to what a smart factory is. Descriptions compiled from industry studies and reports define smart factories by:

- Significant improvement in the development of manufacturing businesses and their position in the supply chain.
- Completely connected and flexible systems, which rely on the constant data flow from connected production and operations systems.
- Beyond typical automation in a production facility, especially in terms of executing discrete tasks or processes.

These rather fulsome descriptions help but only a little. In a very general way, smart factories are the focus of resources, processes, and practices. In an Industry 4.0 environment, smart factories are where manufacturers generate value (more about that later).

5.1.1 Characteristics of a Smart Factory

Because there's no accepted definition of a smart factory, it's difficult to ascribe "typical" characteristics to it. Instead, it's easier to get a clear picture of what a smart factory is by describing what it can do. This smart factory indicator approach provides these capabilities:

- Monitors, collects, coordinates, controls, and integrates data by using IT communications and data management technology.
- Produces and distributes manufactured goods more quickly in response to market demand.
- Uses intelligent agents and other cyber-physical systems to operate more efficiently.
- Uses automated agents to optimise complex production decisions typically relegated to humans.
- Uses digital connectivity to collaborate with suppliers, customers, partners, and departments within the facility.

- Can be connected to a global network of similar production systems and the digital supply chain.
- Evolves to meet the changing business goals of the organisation.
- Adapts to and learns from new and changing conditions in real-time or nearly real-time.

Cyber-physical systems (CPS) play a critical role in this constant churning of technology and process change. They add new capabilities to physical systems by merging computing and communication capabilities with physical processes. The power of CPS lies in their ability to provide new capabilities that enable increasingly complex manufacturing processes.

5.2   What Makes a Smart Factory "Smart?"

That lack of consensus in what a smart factory is extends to what makes it smart. However, those who hope for a consensus might find that one is on the way. Recent studies that assign smart factories with specific design and production capabilities are making their way to the manufacturing trade press and consultancy white papers.

5.2.1   Smart Factory Design Principles

'Industry 4.0: The Fourth Industrial Revolution', a study by the I-Scoop consultancy, describes the smart factory indirectly by suggesting functional requirements that each smart factory should include or use. These requirements, which emphasise what is possible today with existing technology, are based on six basic factory design principles.

Modularity — This design capability enables system components to be assembled, disassembled, and recombined quickly and easily. On the production floor, this translates into being able to add, relocate, or rearrange components in the production line with minimal time and effort. A highly modular smart factory design enables the rapid integration of smart assets, which can be supplied by multiple vendors.

Interoperability — A smart factory design that emphasises interoperability ensures that technical information can be shared within [or between] system components. Such business information can then be shared between manufacturing enterprises, suppliers, and customers.

Decentralisation — Bringing decentralised and autonomous decision making to machines and cyber-physical systems is a core goal of Industry 4.0. The focus is on autonomous system elements, such as modules, material handling systems, and products located anywhere on the production floor. The general goal is to enable CPS to make decisions without regulation by centralised control (man or machine).

Here are two possible standards of autonomy in smart factories:
- Enable CPS to make production process decisions autonomously in real-time, if the outcome does not violate high-level business goals.
- Let embedded computers help autonomous cyber-physical systems interact with their production environment via sensors and actuators.

Real-time capability – Based on the modularity the smart factory should be configured / self-configured to respond to the change -both internal and external – on time. This fast response is based on the capability of collecting and analysing up-to-date data. With the capability, manufacturers will gain the insights of root-cause and predict potential risk of unplanned shutdown, as well as schedule the production line shift based on the ever changing customer demand.

Virtualisation — This process combines physical manufacturing systems, their digital equivalents, and process data to create a virtual factory environment. In this virtual environment, it's possible to:

- Monitor, control, and simulate physical systems and processes.
- Send data to update the virtual model in real-time.
- Make design changes to the factory by creating digital prototypes.
- Train the workforce to perform manual tasks.
- Diagnose and predict faults.
- Guide employees in maintenance tasks.

Service orientation — This design principle shifts the focus from selling products to selling products and services. Smart factories with a service orientation strategy will design and produce products, create related services, and sell them together. This approach encourages the innovative improvement of core processes and if necessary, the outsourcing or elimination of other processes.

- Responsiveness — This essential capability reacts to changes in the status of internal production systems, customer tastes, or other changes in the market. Responsive smart factory designs:
- Use real-time data monitoring and analytics methods to identify process, equipment, or market changes.
- Include enough modularity to expedite system recovery or changes to production processes or equipment.
- Include real-time responses to internal changes, monitoring, and control.

5.2.2 Requirements for a Smart Factory

How do these basic design recommendations translate into functional and physical specifications for a smart factory? The pioneering researchers and practitioners do provide guidelines for industry's reference. A smart factory on earth is a manufacturing facility – highly digitized and fully connected though. Hence some basic elements will form the base to the future-proven production.

Integrated data and processes — The requirement refers to both vertical and horizontal integration. As smart factory is the heart of Industry 4.0, both information technologies (IT) and manufacturing technologies (OT) will be implemented across the geographically disparate sites and between the manufacturing company and other firms across the entire value chain. By horizontally integrating various information systems, the manufacturer will be able to make smooth transition to cyber-physical system based production with deep interconnection and data exchange. Regarding vertical integration, the Industrial IOT enables the instantaneous access to both IT and production systems. Through the framework named "digital thread" data and information is directly extracted from the shop floor by controlling devices and sensors way up to the corporate board room. Data is appropriately processed at various hierarchy levels through production, automation, operations management and corporate management.

Multi-skilled workforce —Smart factory has created an unprecedented operations list with much increased automation never witnessed before. The workers on today's shop floor have to face the fact that their tasks will likely disappear in near future, replaced with new ground and new rules. These can include decision making, supervision, maintenance, programming, or performing a collaborative task or process together with robots. This new paradigm requires the new generation of workers to look beyond the current job scope, and take into account the bigger picture of enterprise, partner eco-system and end user satisfaction. In addition, new responsibilities bring up the new opportunities of lifelong learning to both working staff and corporate management. All these elements will form the base to support sustainable operations for future manufacturers.

Cyber security — Cyber security has never been a new concept since Industry 4.0 initiative was brought in people's eyes. Due to the legacy of manufacturing, Industry 4.0 changes the cybersecurity landscape. On the other hand cybersecurity has become a main enabler of Industry 4.0. A fully connected factory with smart devices and smart networks can be exposed under major attack risk, and requires robust solutions to protect hyper connected systems from unauthorized access and damages as potential consequences. In this regards, cyber security has become a part of corporate strategy to focus on and commit to that supports the sophisticated functions in smart factory operations.

Smart product — Going beyond the classic definition, a smart product stands for a data collection and processing tool with interactive functions. A smart product provides all information required for the production processes, identifying itself to the available modules in the current environment. For example, RFIDs are attached to the product to make product-related information recognisable for and accessible to warehouse inspectors. Thereby shortening operation times tremendously. Smart products nurture and expedite the new business models by providing data-driven services, and create intelligent cross-selling & up-selling opportunities for a smart factory.

After-sale services — This requirement reflects the expanded concept of manufacturing in Industry 4.0. Since after-sales service is shifting from a purely transactional model to service level guarantee model (such as SLA or subscription based service), after a product is built and sold, the value chain won't be complete until manufacturers offer and sell related services. Enabled by the advanced manufacturing technologies, manufacturing companies will be able to deliver predictive maintenance, hence improve the product uptime and profitability.

5.2.3 Drivers & Enablers of Smart Factory

The push toward adoption of smart factories and Industry 4.0 approaches to manufacturing is a reaction to these business trends.

- An increasingly complex global supply chain
- Global fragmentation of demand and production
- Increasing pressure from competitors and unexpected sources
- Constant labour challenges

Several important technology enablers complement these trends:

- More powerful computing and analytical capabilities.
- Newly developed "stems" of smart assets.
- Cloud-based data storage and management services.

When used together in a product design-to-customer environment, these advances enable smart factory processes, which learn from real-time production, logistics, and marketing data. This capability allows smart factories to operate in a more proactive, responsive, and predictable manner.

6 Connectivity & Interoperability: The Ultimate Enablers

"It is time to re-imagine how life is organized on Earth. We're accelerating into a future shaped less by countries than by connectivity. Mankind has a new maxim – Connectivity is destiny – and the most connected powers, and people, will win."

— Parag Khanna, Connectography: Mapping the Future of Global Civilization

One hidden element underpins Industry 4.0 and the entire 4th Industrial Revolution (4IR). Without it, very few of the recent and future transformative digital technologies we discuss would be possible. This emerging technological era has been dubbed 'the information age', 'the data age,' and the cyber-physical revolution, but an equally accurate description would be - The Age of Connectivity.

Connectivity is at the heart of Industry 4.0, enabling disruptive mobility by providing accurate real-time information and unprecedented freedom to many roles within the industrial and business landscape. Connectivity facilitates cloud computing, which is bringing the power of big data analytics and artificial intelligence to companies of all sizes.

Harnessing connectivity is the key to developing automation, robotics, virtual and augmented reality, Blockchain, and almost all Industry 4.0 systems. Without connectivity, data would be a blessing and a burden suffered by the wealthiest companies, and only by increasing connectivity will we see the fourth industrial revolution (4IR) truly blossom.

However, connectivity is useless unless the machines communicating can understand one another. So, in our journey to Industry 4.0, the field of interoperability has come to the fore to facilitate the smooth exchange of information between various devices and systems, often made by a range of manufacturers.

One of the good examples for addressing the criticality of connectivity is Singapore – a brilliant achiever in various endeavours.

"Our vision is for Singapore to have a thriving Digital Economy, where every business is a digital business, every worker is empowered by tech, and every citizen a connected citizen," states the Infocomm Media Development Authority (IMDA) in a May 2019 consultation paper. "World-class connectivity infrastructure will be essential to achieving these objectives," the document continued.

The Singapore Government's Smart Industry Readiness Index (SIRI) lists connectivity as one of its three technology pillars, the index stating that:

Connectivity — measures the state of interconnectedness between equipment, machines, and computer-based systems to enable communication and data exchange across assets. IoT-enabled devices are also increasing in both quality and quantity, generating enormous amounts of data as a result.

Technological advancements in cloud computing and wireless infrastructure also make it possible for data to be centrally collected and managed. Likewise, systems that were once independent or isolated can now be integrated, unifying the various shop floor, facility, and enterprise systems through connected organisation-wide networks.

Interoperability — the ability to access data with ease across assets and systems, is key to achieving this [connectivity]. Companies need to standardise or make use of complementary communication technologies and protocols to establish more open, inclusive, and transparent communications networks.

6.1 Creating Connectivity

Connectivity, in a digital sense, is defined by the Oxford dictionary as the "capacity for the interconnection of platforms, systems, and applications." Connectivity, therefore, includes every machine to machine interaction be it via traditional telephone and Ethernet cables, Wi-Fi and cellular networks, or the latest industrial communication protocols.

The concept of connectivity in the context of 4IR, however, generally refers to it as the capacity for different nodes of a network to communicate with one another. More connectivity means more nodes and increased flow of data, which in turn leads to the greater intelligence of systems. Only by increasing connectivity can we hope to develop the fully automated factories, distributed renewable energy, cognitive buildings, and smart cities that have begun to define our future.

The primary role of connectivity in Industry 4.0 is to enable companies throughout the manufacturing supply chain to form networks and optimise individual steps in the supply chain. Various information and communications technologies enable creation of networks, which include entire manufacturing processes. Smart sensors and devices connect warehousing systems, machines, human workers, and production operations together to bring about a wide range of enhanced processes and services.

6.1.1 Connectivity Today

Our quest for connectivity can be traced back long before the first telephone, to the mail and messenger services that preceded it. However, in this digital age of smartphones, the internet, and process automation, the forefront of connectivity in manufacturing today is a broad range of cellular and industrial network protocols. Each racing to develop the characteristics that will make it a leading connectivity platform for Industry 4.0.

Industrial protocols are the cornerstone of interoperability in the manufacturing facility, developed to interconnect the systems, interfaces, and instruments that make up an industrial control system — often deployed throughout broad network architecture, including business, plant, networks, or Fieldbus networks. Popular industrial network protocols include Ethernet/IP, Ethercat, Modbus, and IO Link, among others, each offering different characteristics to suit their application.

Ethernet has long been seen as an ideal connectivity solution for industrial network communications due to being an open, proven, cost-effective, world-wide standard that's easy to implement and use. The 100 to 1000 megabit per second data rates it supports are significantly higher than most existing industrial field buses. IP adds integration and data transparency on all networking levels, allowing a seamless flow of data from the factory floor to the back office for management and control.

Ethernet/IP, and other industrial network protocols, however, are not sufficient to support the complexities of Industry 4.0 networks. Even though standard Ethernet protocols define communications from the physical hardware layer to the communications application layer of a network, they do not include user application levels, such as data formatting to enable data exchange between equipment.

SigFox and LoRa have long been the major players in the Low Power Wide Area Network (LPWAN) space, each offering low cost, low energy consumption, broad coverage, and significant capacity. However, their bandwidth per object is very limited (100 bps), and the latency is high (1s), preventing their use in deployment of the real-time applications that symbolise Industry 4.0. The 868 MHz and 920 MHz wavelengths used also poses challenges indoors, meaning many manufacturing facilities have sought alternative connectivity solutions.

Cellular networks are by far the most widely used platforms digital connectivity today. 2G, or GSM (Global System for Mobile), began deployment in 1992 as the first all-digital cellular communication standard. Despite the subsequent emergence of 3G and 4G, the older 2G-GSM is the dominant global cellular standard today with 80% of the market share and close to 5 billion subscribers. Even as the IIoT emerges, 2G-GSM remains an effective way to facilitate some of the machine-to-machine (M2M) connectivity that enables Industry 4.0.

Many IIoT applications require low-bandwidth, low latency, and low energy consumption, so 2G remains attractive due to its maturity. 2G is available widely and accessible by many devices, offering high-levels of integration and low cost of instalment. For applications that send infrequent and untimed data packets, such as automatic meter reading (AMR), signage, and some types of sensor data, 2G is just as capable as its successors but at a fraction of the cost.

The key challenges facing 2G are its increasingly limited range of applications, as Industry 4.0 increases bandwidth and latency demands, but also because it is being slowly phased out across the world due to the maturing of 3G and 4G / LTE technology. Singapore, for example, decommissioned its 2G network in 2017. While major global operators, such as AT&T, Telstra, Optus, and Vodafone, have gradually been shutting down their 2G services in some territories. It only a matter of time until the same trend takes shape in all ASEAN members.

3G is an evolution of the 2G-GSM communication standard, launched in 2000. 3G offered data transfer speeds of up to 14 Mbps (more using packet switching), four times faster than 2G. It provides greater clarity for video-streaming and real-time communication by using a Wide Band Wireless Network with a range of 2100MHz and has a bandwidth of 15-20MHz. 4G offered similar features to 3G but significantly enhanced it with better use of bandwidth and speeds of 10Mbps-1Gbps.

Mark Hung, an analyst at research firm Gartner, points out that "3G brought web browsing and data communication to mobile devices, 4G greatly enhanced it. And even though towers today can support

71

hundreds or thousands of devices, 5G could help scale the Internet of Things from hundreds and thousands to hundreds of thousands."

Considering that many, including tech-giant CISCO, predicting that the number of connected devices on the Internet could exceed 50 billion by 2020, the deployment of 5G cannot come soon enough.

6.1.2   5G & Industry 4.0

3G to 4G was a small step for connectivity, but 4G to 5G will be a giant leap for Industry 4.0. Adoption of 5G won't just make connectivity faster and broader; it will trigger the dawn of a new age of data-intensive 4IR technologies, such as driverless vehicles, intelligent robots, wearable and implanted technology, virtual and augmented reality, and many more that promise disruptive industrial applications.

While 4G-based Industry 4.0 pilot projects are emerging, 5G will unleash an unprecedented scale of connectivity. 5G is promising to be the connectivity leap that will make Industry 4.0 technologies available to the mass market.

Andrew Ross, an analyst for the Information Age, believes that with the next generation of the industrial revolution being triggered by the combination of emerging technology, the impact that 5G has on Industry 4.0 will be unique. As a trend itself, 5G won't redesign the production line, but it will enable new operating models. Furthermore, with network characteristics that are essential for manufacturing, 5G will offer manufacturers the chance to build smart factories that can take advantage of the emerging tech that's changing the industry.

5G brings robust, high-speed data transfer through very low (1ms) latency to enable the dependable, real-time, data-rich connectivity these advanced technologies demand. 5G offers slower speeds too, designed to increase range, and low power consumption, helping the IoT take hold in large, remote, or distributed deployments.

The sheer volume of data that can be transferred over 5G networks will bring powerful artificial intelligence (AI), machine learning, and big data analytics, which will drive efficiency and productivity to new levels. Figure 6-1: 5G Key Capabilities highlights the key benefits of 5G over 4G in a range of 4IR applications. When analysts and manufacturers think about enablers of Industry 4.0, their minds often go straight to the standard technology enablers such as artificial intelligence, virtual reality, or the Internet of Things.

It's easy to forget that connectivity —the function as well as the technology— is the ultimate enabler of digital manufacturing. If the ASEAN region hopes to compete in the Industry 4.0 era, upgrading connectivity infrastructure is an essential element.

The IMDA's Second Public Consultation on 5G Mobile Services and Networks states that 5G, globally acknowledged to be the next big leap in mobile and wireless communications, will be a critical part of this infrastructure. More than higher speeds, 5G will enable more things to be connected, with better reliability and lower latencies.

The document continues with the statement that the 5G network architecture will allow operators the ability to customise and tailor services to meet the demands of different end-users, to support the innovative services and applications driving Singapore's Digital Economy.

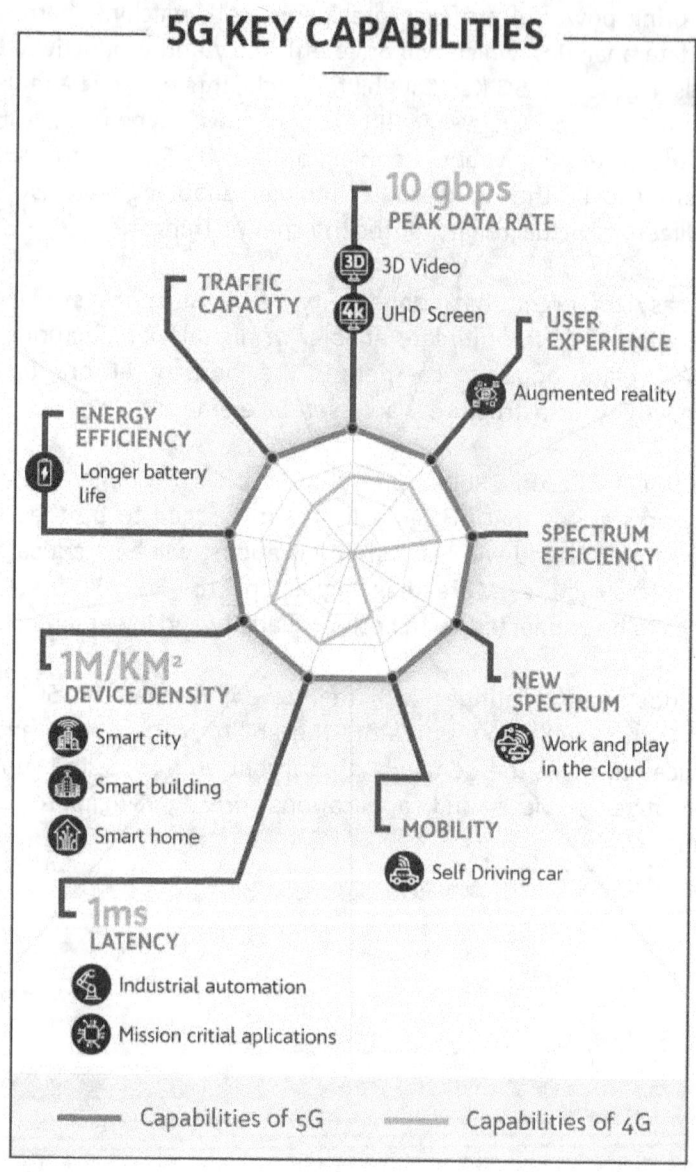

Figure 6-1: 5G key capabilities

74

The emergence of 5G is not without friction, however. Strong criticism and difficult challenges persist. In early 2019, a petition opposing the roll-out of 5G was signed by over 26,000 scientists on the grounds of health and environmental concern. Others in the scientific community dismiss these claims and point to previous generations of cellular connectivity.

Further challenges, like spectrum bandwidth conflict with strategic infrastructures, such as weather satellite communication, represent technical hurdles to overcome. While complex regulatory demands also threaten to delay ambitious timelines set by many markets.

Several ASEAN members will have to consider significant regulatory reform before they can start their full 5G roll-out. Previous iterations may have succeeded but 5G is a different ball game with broad applications - all actors will need to work together to bring it to fruition, especially in the manufacturing space.

The Singapore 5G public consultation document states that besides commercial readiness of 5G technology, IMDA will have to take other factors into consideration when allocating spectrum for 5G services. These include the harmonisation of 5G spectrum bands amongst neighbouring countries in the region, refarming of spectrum frequencies assigned for existing use and re-assigning them for 5G uses, and addressing coexistence issues with the neighbouring countries, amongst others.

International industry bodies hold critical roles in the development of 5G. The 3GPP (Third Generation Partnership Project) is focused on developing high-reliability and ultra-low latency radio technologies and architectural components that can support the Industry 4.0. The 5G Alliance for Connected Industries and Automation (5G-ACIA) and the EU's 5G Infrastructure Association (5G-IA) meanwhile, serve as global forums to address, discuss, and evaluate challenges and opportunities.

Major ASEAN 5G and Industry 4.0 stakeholders must also come together to ensure smooth 5G development in the region.

6.1.3   5G in the ASEAN Region

ASEAN region and across the world. Lured by faster data transfer speeds, greater bandwidth, and more connections, all ASEAN nations have announced some intention to bring 5G connectivity to their country. The diversity of ICT infrastructure standards, as well as socio-economic and political capacity to upgrade them, will mean a staggered introduction of 5G across the region.

The promise of Industry 4.0 represents a driving force for the region's 5G development. Manufacturing is a crucial avenue of economic growth for many ASEAN members; the sector contributed 21% (approximately $670 billion) to the region's total GDP in 2018, which is predicted to double to $1.4 trillion by 2028.

According to global management consulting firm, AT Kearney, ASEAN will have about $250 billion to $275 billion in incremental value at stake by 2028, representing a 40% increase in manufacturing value added (MVA), via productivity gains and new revenue streams offered by 4IR technologies. There is little doubt that 5G will be fundamental to the future of ASEAN manufacturing.

The AT Kearney report states that the rise of Industry 4.0 technologies poses a major threat to ASEAN manufacturing and its growth potential. Manufacturers' low-cost competitive advantage is gradually being eroded as competitors in advanced economies use new technologies to achieve significant improvements in cost, speed, quality, and sustainability.

Meanwhile, contrary to the threat, Industry 4.0 also presents regional manufacturers that embrace the digital revolution with a significant opportunity: the ability to leapfrog onto the global manufacturing stage. It is therefore crucial for ASEAN manufacturers to accelerate their 4IR adoption or risk being left behind.

Singapore is the regional leader for connectivity. Its modern industry, successful smart technology initiatives, and highly-educated population mean the economically prosperous city-state offers an ideal environment for 5G. Singapore is not alone; however, other ambitious ASEAN members proved their abilities with successful 4G deployments and have announced strong intentions for their 5G roll-out.

Figure 6-2: The Impact of 5G in various verticals and use cases

According to a recent report by London-based Opensignal entitled 'The State of Mobile Network Experience,' Singapore performed best for 4G provision, followed by Thailand, and Indonesia showing that cellular connectivity can be successful in a variety of socio-economic and political environments.

The report, published in May 2019, notes that 4G availability shows the proportion of time Opensignal users with a 4G device have a 4G connection. However, this is not a measure of coverage, and therefore does not account for some of the more challenging Southeast Asian geographies.

Ultimately, each ASEAN member state will have to bring 5G online themselves, but pressure and confidence will rise as the region's progressive nations begin deploying. However, the value of greater cooperation between all stakeholders will also be critical to elevating connectivity and enabling Industry 4.0 across the whole region.

Maziar Nekovee, Professor of Telecommunication, 5G Mobile Technologies, and Head of Department of Engineering and Product Design at The University of Sussex, UK suggested that the key to achieving the enormous potentials of the marriage of Industry 4.0 and 5G is collaboration between stakeholders from the manufacturing and mobile industry ecosystems, which in the past, have been largely operating in parallel. He also mentioned that, "Traditionally, the focus of the mobile industry has been the provision of conventional service, voice, video, and data to consumers while the manufacturing industry has been relying on its own solutions, or those retrofitted from the IEEE family of wireless technologies to support limited connectivity inside factories, plants, and warehouses. With 5G all of this is changing rapidly."

6.2 Enabling Interoperability

Interoperability can be defined as the ability of two or more products, programs, or systems, to exchange and interpret each other's data. Ability, in this sense, can be referred to both the means to interact and the quality of that interaction. Connectivity allows the conversation to take place between these elements; interoperability ensures they speak the same language.

The factory, and the wider manufacturing supply chain, has evolved through successive industrial eras into an incredibly diverse technology landscape. Every production facility is made up of a plethora of equipment, of different ages, from numerous manufacturers, and with varying approaches to communication.

Industry 4.0 wants to connect all those "things" and more, but only if they can talk to one another. To bring about the vast benefits of a smart connected factory, manufacturing operations will need to enable a high level of interoperability.

Interoperability has become a key challenge for the development of Industry 4.0 technologies. The sector has been slow to create and agree upon adequate standards for secure exchange of data between machines, devices, and services across different companies and industries, leaving us with a somewhat fragmented interoperability landscape.

However, as we move into a new decade there is optimism about the increasing collaboration within the industry and the interoperability standards that are currently establishing themselves in the manufacturing space.

6.2.1 Interoperability Standards

The whole Industry 4.0 initiative focuses on connecting IT and automation in a strategic way, thus forcing interoperability technologies such as Open Platform Communications Unified Architecture (OPC UA) in the spotlight. Industry 4.0 and related initiatives accept that open software and communications standards are vital to self-managing production processes.

In order to drive the Industrial movement and bring about the economic growth it promises, standards are essential to enable controllers, sensors, equipment, machines, humans, and other systems to cooperate and communicate directly.

Design engineers of Industry 4.0 demonstration processes acknowledge this need by using existing standards, such as the ISA95 integration standards (discussed in 3.3.2) in their designs. Adopting these standards recognises that manufacturing processes are filled with international interdependencies, which require shared standards and interoperable technologies. Hence, Singapore Smart Industry Readiness Index (SIRI) encourages companies to standardise or make use of complementary communication technologies and protocols to establish more open, inclusive, and transparent communications networks.

Three standards have emerged as the most important building blocks of Industry 4.0 interoperability, each with its own history, best use cases, and capabilities within the manufacturing supply chain. Critically, these standards are platform independent, which enable devices sold by different vendors to exchange information without system hang-ups.

Open Platform Communications (OPC) is the prominent interoperability standard, and a vital element in digital manufacturing and automation efforts. OPC is an interoperability standard used to secure the exchange of manufacturing automation data.

Early versions of OPC (or OPC Classic) —OLE for Process Control— were plagued by security problems, closed architectures, maintenance difficulties, and time-consuming data management. Its complexity and expense made it difficult to run and unsuitable for use in Industry 4.0 applications.

Early Industry 4.0 adopters would need a standard that enables improved functionality and usability, OPC responded with the significant Industry 4.0 improvements built into the updated version — the OPC UA standard.

Features of OPC UA include:
- **Machine-to-Business** — OPC UA is the first communication technology built specifically to run in the grey zone, where data must cross firewalls, security barriers, and specialised

platforms to make that all-important, machine-to-business connection.

- **Efficient Data Handling** — OPC UA provides a very flexible and adaptable way to move data between enterprise IT systems and the controls, sensors, and monitoring devices that handle real-world data.
- **Tool Utilisation** — The OPC UA protocol connects analytic tools, databases, enterprise resource planning (ERP) software, and other IT solutions to real-world data collected from low-end controllers, sensors, actuators, and monitoring devices.
- **Backward-Compatible** — Based on extensible SOA(Service Oriented Architecture) framework the OPC UA is backward-compatible with OPC and uses a software design approach that has become widely used in manufacturing systems around the world.
- **Platform Availability** — OPC UA supports a wide range of platforms, from embedded microcontrollers to cloud infrastructure for greatest integration flexibility.
- **Enhanced Security** — The OPC UA standard provides high levels of security through encryption, authentication, and auditing.

Based on the statement of OPC Foundation, the vision of IoT can only be realised, if the underlying communication between central components is based on a global communication standard that can fulfil a wide range of complex requirements. Thus, using a common standard makes a lot more sense to preserve the value of overall datasets.

OPC UA ensures the open connectivity, interoperability, security, and reliability of industrial automation devices and systems. Currently, OPC UA has become the main communication and data modelling standard for Industry 4.0. It's completely scalable, works with many software platforms, and its widespread adoption is quickly making OPC UA the backbone of Industry 4.0 connectivity.

6.3 Getting Connected to Industry 4.0

The wealth of benefits and its futuristic aura has led many small- and medium- sized manufacturers to believe that Industry 4.0 is an expensive endeavour reserved for the major players. This is not the case, however. Industry 4.0 technologies and systems are inherently scalable, designed to grow with companies. Most applications are focused on cost-savings and increased productivity, allowing firms to do more with less and promising fast ROI.

ASEAN focused research suggests a gradually growing awareness in the region that a "low-and-slow" approach to Industry 4.0 can suit companies of all sizes, although trade media paints a picture of shiney, 100% digital, highly-connected, fully-automated factories.

The truth is that Industry 4.0 is increasingly being used to supercharge legacy equipment and established operations, application by application according to needs and budgets. All enabled through connectivity and interoperability.

Those looking to begin their path to Industry 4.0 slowly have a wide variety of options to suit their unique situation. There are opportunities to connect existing and new manufacturing technology, for example. Even old machinery can be connected to monitoring software indirectly with a simple sensor, albeit with a minor loss of accuracy. These can provide initial AI and automation applications, which can be built upon gradually as the benefits are demonstrated and the overriding strategy develops.

Paul Miller, senior analyst at Forrester addressed that a lot of industrial machines have been putting out some sort of electronic signals, but connecting them with the modern analytical platforms is certainly a tough job to do. He suggests manufacturers to set right expectation on the Improvement that single step can provide, and a smart factory can simply be a prototype sensor measuring things since all that really matters is that the factory technology connects.

Consider this simple 4-step strategy for a manufacturer to launch their Industry 4.0 ambitions:

1. Upgrade to OPC-UA IIoT architectures.
2. Tie the architecture into an existing factory automation infrastructure.
3. Feed factory data to the latest cloud-based advanced analytic tools.
4. Use data analytics and visualisation tools to monitor and optimise processes in line with company goals.

With the right guidance, digital transformation doesn't need to be a stressful and disruptive process. As always with technological transformations, early adopters take the risk to earn substantial rewards.

In Industry 4.0, however, the scalability of connectivity allows manufacturers to "test the water" with a range of approaches to suit their needs. So they can understand what Industry 4.0 can do for them before they decide how to revolutionise their business.

Vidya Ramnath, Vice President, Global Plantweb Solutions & Services, Emerson Automation Solutions, states that one of the key disruptive forces of Industry 4.0 is the ever-increasing volume, velocity and value of data. Looking ahead, traditional manufacturing companies need to change their perception of data, not just as numbers on a screen, but as a strategic asset that can unlock revenue growth and deliver cost savings. Companies who embrace this shift will start building the infrastructure for connectivity and intelligence right away.

7 Setting Up an Industry 4.0 Practice

"Knowledge is a treasure, but practice is the key to it."
Lao Tzu.

Getting beyond the exciting generalities of Industry 4.0 theory can be difficult. Part of the problem is that even early-adopter results are difficult to find. For every pound of digital manufacturing vision and capabilities documents that one reads or hears about, a meagre ounce of practical experience and advice appears for review.

We hope to fill this information gap by providing a gradual, budget-friendly approach to starting an Industry 4.0 practice. Then, we'll share the progress of ASEAN-6 early adopters that we can find.

7.1 Coming Down To Earth With Practice

There's an accessible approach to starting an Industry 4.0 practice, one that companies of any size can benefit from. First, it's necessary to address some possibly damaging ideas about what it takes to succeed.

7.1.1 Getting Beyond the Myths

There are several myths about Industry 4.0 that can make the most forward-looking manufacturer think twice about adopting its methods and technologies. You might have encountered them amongst all the hype and enthusiasm that's floating in the media and trade press.

Putting these myths to rest is an important part of understanding the benefits and requirements of adoption. Despite what you might have heard, successful Industry 4.0 projects can be:

- **Completed gradually in stages** — There's no need for a Big Bang deployment. Manufacturers often assume that they must replace entire, existing systems. On the contrary. Legacy systems can be upgraded to provide smart manufacturing capabilities. A complete system overhaul is not necessary.
- **Developed quickly as a series of agile projects** — Adding Industry 4.0 capabilities can get tricky, but it doesn't have to be a lengthy process. The idea is to chop the entire modernisation process into manageable pieces. Then it's a matter of developing, testing, retesting, and launching the improved process quickly. This agile approach to modernisation is sometimes known as "fail quickly process development". If the development step fails, you don't risk much in time and treasure to try again.
- **Implemented in phases to control costs** — An incremental approach is not only useful. In many cases it's advisable. No rule requires that everything on a production line to change at once. Project engineers control which infrastructure, equipment, and processes change in Industry 4.0 modernisation. Improvement projects can be any size that technical specialists think are appropriate.
- **Engaged successfully at companies of all sizes** — At this very early stage of digital manufacturing development, there are relatively few reports of Industry 4.0 adoption in SMEs. Perhaps, only larger enterprises get the lion's share of press coverage. Whether or not that's true, SMEs can adopt and benefit from digital manufacturing improvements.

The World Economic Forum study, Fourth Industrial Revolution Beacons of Technology and Innovation in Manufacturing proves this notion. Elettrotecnica Rold Srl is an SME at Cerro Maggiore, Italy. Its 250 full-time employees make washing machine door locks.

The company's factory-applied digital manufacturing technology at scale to increase productivity and product quality. Their experience shows that innovation is possible, even by using off-the-shelf technology and collaborating with universities and local technology providers.

There is more and one way to create   successful Industry 4.0 engagement. While there's no single procedure that works for everyone, it is helpful to assemble a very specific set of information, approaches, attitudes, and relationships, which can be tweaked and moulded for all situations.

7.1.2 A Modest Plan, Executed Time & Again

You could summarise our modular approach to Industry 4.0 practice as:

Optimise — Augment — Test — Launch — Repeat

The general idea of this approach is to start with existing assets, machine and human. Optimise their performance; add new equipment or processes as needed. Finally, test, retest, and launch the improvements, quickly and often. Then, find the next improvement and repeat the entire process.

Much of the transformation in digital manufacturing improvement projects involves changes in the attitude, approach, and focus of early adopters themselves. Industry 4.0 modernisation practices can become an opportunity to view and carry out process improvements in a new way.

Consider these suggestions before you start:
- **Don't be afraid to start small** — Thinking modular is an excellent way to start. We're talking about a series of small, high-priority improvements that you can add to your existing

systems and legacy equipment. The best proof of concept is a modest success, which you can add to as resources and acceptance of the newly improved process grow.

- **Get entrepreneurial** — We're not talking about finding new business models or developing new product designs. You might want to delay these forays into innovation until you've achieved initial success. No, we're talking about making production, quality, and post-production processes more efficient, secure, and safe.

The trick is to find opportunities to practice, optimise, augment, and rapid test-and-launch approach to improving small parts of a facility's capabilities. For example, by connecting all your production, product, and customer data sources and feeding information onto a big data analytics platform, you can avoid data silos and make your supply chain visible from end to end.

There's a lot of ground to cover in your search for improvement opportunities. And important is that the RAMI 4.0 model expands the traditional concept of manufacturing into value chain steps that occur before and after what happens on the shop floor.

- Get clear about goals and performance. It's essential to link your improvements to an important business goal as defined by a critical success factor. Then, define desired results in terms of key performance indicators (KPIs) that your stakeholders' value and trust. This gives you a clear definition of success.
- Get comfortable working in an agile environment. Early success stories of Industry 4.0 adoption indicate that small projects that are tested and launched quickly and often are most likely to succeed.

Currently, these principles enable successful Industry 4.0 deployments in Southeast Asia and beyond. To learn more about our process we provide a deeper dive with details and examples of useful data next.

7.1.3 Detailed Industry 4.0 Start-up Process

It is all too common for Industry 4.0 early adopters to drown in an ocean of vision statements and good intentions with little down-to-earth "how-to" support to guide them. By using a gradual, modular approach to Industry 4.0 adoption, however, it's possible to achieve economic benefits quickly, even on a modest budget.

Step 1: Identify an opportunity to improve operations — try to look for ways to achieve economic benefits quickly, i.e., industry companies are encouraged to be concentrating on short-term operational improvements rather than new products or business models, finding a potential process improvement that can involve production stage processes or ones that come before (design/prototyping) or after (distribution, after-sales services) the manufacturing process itself. Just make sure that the improvement reflects a high-priority business goal.

Step 2: Identify the change story — come with the story of the changes the management team wants to make. It's worthwhile to use 25 words or less to describe what you want to happen. Here are a few examples taken from a recent research paper:

- Highly advanced factory with lean processes uses digital manufacturing to reach the next level of operations performance.
- Fourth Industrial Revolution technology use cases target quality improvement and cost reduction to meet customer expectations.
- Staff of 50-year old plant recognises that they must use digital tools to stay price-competitive for the next 50 years.

- Use big data from connected machines to improve operations and use agile proof-of-concept to support rapid deployment of new use cases. These stories describe high-level intentions.

Next, it's time to specify the situations that help define the nature and scope of your process improvement.

Step 3: Identify use cases — List the situations or process functions that are most likely to deliver the results that company wants - such as digital inventory management, for example. Don't forget to use cases that enable scaling. Use cases are the bridge between the intentions and implementation.

Here are the emerging technologies use cases related to the business operations with value:

- IIoT-enabled machine condition monitoring
- Digital inventory management
- Digital tool life-cycle management
- Digital value stream mapping
- Real-time labour processing and time tracking

The trick here is to decide what your problem is, the best way to solve it, and which use cases should be part of the solution. This requires a clear definition of the specific targets the digital team wants to achieve.

Step 4: Establish an Information Technology/Operational Technology (IT/OT) infrastructure — Even if the team is engaging in a modest pilot, it's part of a step change transformation. Before any compay can start the transformation, the team must prepare their own IT/OT infrastructure in several critical areas:

- Scalable, reliable connectivity
- Intelligent, end-to-end cybersecurity

- An Industrial Internet of Things (IIoT) platform

These tasks involve a lot of infrastructure setup and decision making. The team will be connecting sensors, actuators, equipment, machines, business applications, and data stores throughout the system used in the experiment. The RAMI 4.0 model can help the team define the scope of the connections.

Consider whether you as the planner in a manufacturing conglomerate want to set up your own cybersecurity network or engage a cybersecurity prevention and mitigation service. Then decide whether to manage your data onsite or in the cloud. Industry 4.0 principles assume that your data will reside in the cloud. Nevertheless, some manufacturers report that they don't need off-premises storage, at least not in the early stages of adoption.

Acquire Industry 4.0 products and services — This step involves choosing and deploying emerging technology-related products and services, which enable process improvements. Specific choices will depend on company- and use-specific criteria. However, this is the step in which standards can play an important role in eventual success.

Step 5:Standards for smart factory equipment and services continue to evolve—very slowly. If standards conflicts haven't been resolved when you want to start process improvements, what should users do? It might take years before relevant stakeholders agree about interoperability standards.
Consider partnering up with solution and technology providers — make them prove their value. Expect them to provide relevant use-case proof points at key stages in testing and solution decision making processes.

Expect platform and application providers to offer integrated and interoperable solutions that fit your system's requirements. Make sure that IT enterprise platforms and applications emphasise flexible operation and can use their ERP entry point to expand easily into the OT of your systems.

One thing to highlight - don't forget the big picture. Pursuing modest successes is a good way to begin your digital manufacturing journey, but remember that you deliberately chose these early ventures as low-hanging fruit.

Successful scale-up of Industry 4.0 processes should adopt a two-tier approach, which blends short-term use-cases and a longer-term road map. This approach solves problems at specific pain points and pays attention to clear definitions of performance targets and the development of a strong system of vendors, suppliers, and partners.

7.2 Industry 4.0 Adoption in ASEAN

It's still early days for Industry 4.0 adoption status reports. However, early adopters throughout Asia, Europe, and North America are starting to reap economic benefits from their digital manufacturing projects. Research studies of these early efforts can provide real-life guidance for Industry 4.0 novices.

So, what's happening in ASEAN nations? Most ASEAN-6 manufacturers are still deciding whether to adopt Industry 4.0 or in the planning stage of their modernisation efforts. So, the best course of action is to report early efforts of ASEAN-based manufacturers in terms of their change stories and technology use cases.

7.2.1 Industry 4.0 Change Stories

By expanding our 25-word change story approach, it's possible to summarise a business challenge and potential enablers that ASEAN-6 businesses use in their digital manufacturing projects. Here are several stories that encapsulate three different types of ASEAN digital transformation:

- A Thai garment manufacturer reduces product development times by using IIoT and big data analytics to collect, store, and analyse production data.
- An Indonesia-based electronics manufacturer uses cobots to avoid heavy dependence on manual processes. Their goal: improve process productivity and the quality of their products.
- Malaysian automation specialist offers process intelligence solutions that use big data analytics and AI. The result: faster deliveries and cost savings enabled by fewer waste materials.

These before-and-after stories are precursors of the next step in digital manufacturing development — readiness analysis.

94

8 Assessing Industry 4.0 Readiness

"If you wait for the mango fruits to fall, you'd be wasting your time while others are learning how to climb the tree"

— Michael Bassey Johnson, Master of Maxims

Stories from early adopters tend to agree. One of the most challenging aspects of Industry 4.0 adoption is knowing where you are before you begin. Readiness assessments provide a toolkit, which enables Industry 4.0 adopters to get a clear view of their capabilities and gaps in their IT (Information Technology), OT (Operational Technology), and human resources.

The best bet is to start with familiar methods; gap analysis, benchmarking, and a structured system of indicators. With these tools, you can define your starting point and build a detailed resource inventory that supports your Industry 4.0 development efforts.

8.1 What is a Readiness Assessment?

In old-fashioned project management terms, a readiness assessment is a pre-project review. With it, you measure your organisation's ability to begin a project and identify areas that need more attention before the project commences.

Much of the usefulness of an Industry 4.0 readiness assessment is represented by what seems like a simple question: Is your manufacturing business ready to generate business value by using Industry 4.0 technologies and practices?

8.1.1 One Big Comparison

By comparing manufacturing capabilities, resources, and experience against a set of industry-wide benchmark values, a manufacturing business can use assessment results to create a roadmap. With this guidance, high-level business goals, and performance standards (KPIs), manufacturers can identify and prioritise Industry 4.0 investments and development.

An Industry 4.0 readiness assessment describes the preparedness of a company to undergo a significant change or take on new capabilities, products, or business models. It's also an audit process, which uses checklists of resources and capabilities.

8.1.2 Readiness Assessment Functions

Assessment criteria are based on the current situation at a manufacturing company. For example, an important indicator of Industry 4.0 development success is a company's willingness to alter its current manufacturing, business, and training practices.

By enforcing a first-things-first approach to Industry 4.0 development, assessment results become a useful tool that can be checked and rechecked as Industry 4.0 investments change and grow.

Readiness assessments enforce other helpful disciplines. Measuring company capabilities throughout the supply chain and product life cycle makes it easier for process engineers and business leaders to identify and respond to changes. As they are provided a big picture perspective of their operations and business practices.

8.2 What is Readiness—Ready for What?

In the world of digital manufacturing development, the assets, resources, and methods needed to generate value go beyond the doors of a smart factory. Instead, Industry 4.0 practitioners engage in activities that span the supply and value chains in teams of collaborators within a company and in partnerships that extend beyond it. Readiness involves a lot more than checking a list of capability and resource requirements.

8.2.1 An Approach to Down-to-Earth Improvements

Industry 4.0 practices seldom involve the wholesale replacement of technology and processes. Rarely, a company such as BMW has USD1.4 billion to build a smart factory from scratch. Most businesses, however, must apply more modest, gradual changes. For example, enterprises, especially SMEs, are more likely to add 3D printing for prototyping as an early step in Industry 4.0 development. Or, they might add IIoT sensors and other assembly line devices to network connectivity.

A modular approach to process improvements — This gradual approach encourages process engineers to take a modular approach to improving processes, products, and assembly lines rather than an entire factory.

This, in turn, makes Industry 4.0 readiness a study of both technology and business maturity. What is a business doing now, or what resources do they have at various product life cycle levels, that will accelerate or slow down Industry 4.0 development?

So, readiness is the ability to generate profit and recognise opportunity. Industry 4.0 readiness assessments describe capabilities and practices that enable more profitable operations and business practices. These capabilities and practices are the bedrock of Industry 4.0 development. They help Industry 4.0 assessors identify needed assets, resources, practices, and areas of improvement.

The ultimate readiness criterion — Our list of essential Industry 4.0 elements includes resources, as in human resources. Yes, Industry 4.0 assessments include skills development. However, they also involve more subtle aspects of human nature, specifically the ability to change priorities and broaden one's point of view. In study after study, Industry 4.0 researchers include the willingness and ability to change one's business culture as essential to success as adoption of any specific technology.

So, when anyone asks, "Industry 4.0 readiness for what?" think of a one-word answer: "**Change**".

Anatomy of a readiness assessment — Our rough-and-ready description of assessment content —legacy tech and resource requirements as well as maturity indicators and levels— are a start. However, the usefulness of these studies lies in how Industry 4.0 concepts are organised and supported in a standard measurement framework.

The power of readiness assessment lies in the knowledge, assurance, and evidence that a company's proposed changes will be successful if decision-makers elect to proceed. These assessments provide value by:

- Helping to avoid the time, cost, and internal credibility that can be lost when essential resources are not ready to support Industry 4.0 initiatives.
- Enable a company to address potential issues before they become big problems as the initiative moves forward.
- Helping to decide if employees have the skills, resources, and motivation they need to support Industry 4.0 development efforts.
- Identifying areas of improvement before the proposed Industry 4.0 changes begin.
- Recognising an organisation's strengths, especially the things that will serve as strong assets during proposed changes.

A readiness assessment reveals current strengths in resource and asset use and what needs to be improved to maximise value that a manufacturing business can generate.

To provide a high-level view of assessment content and workflow, sections 8.3 and 8.4 provide summaries of two readiness assessment studies. One was designed and tested by German mechanical engineers and manufacturers in 2015. The other is the product of British collaborators: a university, a supply chain consultancy, and a law firm. Each study compares the ideas, procedures, and usability of these different approaches to readiness assessment.

8.3   IMPULS Industry 4.0 Readiness Assessment

In 2015, project partners from the Cologne Institute for Economic Research and the Institute for Industrial Management at Aachen University in Germany developed an online tool. With it, interested manufacturing and engineering companies could measure their own individual Industry 4.0 readiness.

The study that created the assessment was designed to bring the grand vision of Industry 4.0 down to earth. Then as now, there was a need to bridge Industry 4.0 vision and day-to-day business reality. The study authors knew that a useful assessment should:

- Highlight the obstacles and challenges that manufacturers must face during their digital manufacturing development efforts.
- Document the resources and capabilities of mechanical and plant engineering companies before Industry 4.0 development begins.
- Focus on what motivates engineers to change their practices and what holds them back.
- Highlight the different assessment results that occur at SMEs and larger companies.

- Provide engineers and manufacturers with a detailed, systematic framework of Industry 4.0 concepts, capabilities, and maturity criteria.

In 2015, the study authors surveyed about 250 respondents. These manufacturing professionals, managers, and executives tested the model for what later became a readiness self-assessment tool. With it, German engineering and manufacturing companies could measure the capabilities and maturity of their Industry 4.0-related technology, operations, and business practices.

Study framework — The study model begins with the framework, which includes key Industry 4.0 concepts taken from RAMI 4.0 and other foundational Industry 4.0 documents.

The Industrie 4.0 Readiness Assessment model (RAMI 4.0) looks at six levels of capabilities:

- Strategy and organisation
- Smart factory
- Smart operations
- Smart products
- Data-driven services
- Employees

Each of these six levels is divided into 18 fields, which relate to specific operational, training, and business functions.

Impuls/VDMA Readiness Assessment	
Strategy and organisation	Strategy
	Investments
	Innovation management
Smart factory	Digital modelling
	Equipment infrastructure
	Percentage of collected data used in ops
	IT systems
Smart operations	Cloud usage
	IT security
	Autonomous processes
	Information sharing
Smart products	Data analytics in use
	ICT and add-on functionality
Data-driven services	Share of collected data used
	Share of revenue generated with collected data
	Share of revenue generated with data-driven services
Employees	Percentage employees, with no/adequate/high levels of skills
	Percentage employees, whose skills sets have been measured

Table 8-1: IMPULS/VDMA readiness assessment

Source: https://industrie40.vdma.org/documents/4214230/26342484/Industrie_40_Readiness_Study_1

Before users can measure their Industry 4.0 readiness, however, the study framework must be expanded to create a comprehensive system of indicators.

8.3.1 Measuring Industry 4.0 Readiness

The Industrie 4.0 Readiness tool enables users to measure the maturity of their information and operations technology as well as their management and business practices.

The framework arranges these sub-dimensions along a spectrum of capabilities, which measure a company's technology and business maturity. Figure 8-1 is an example taken from the Employees dimension of the Industrie 4.0 Readiness study:

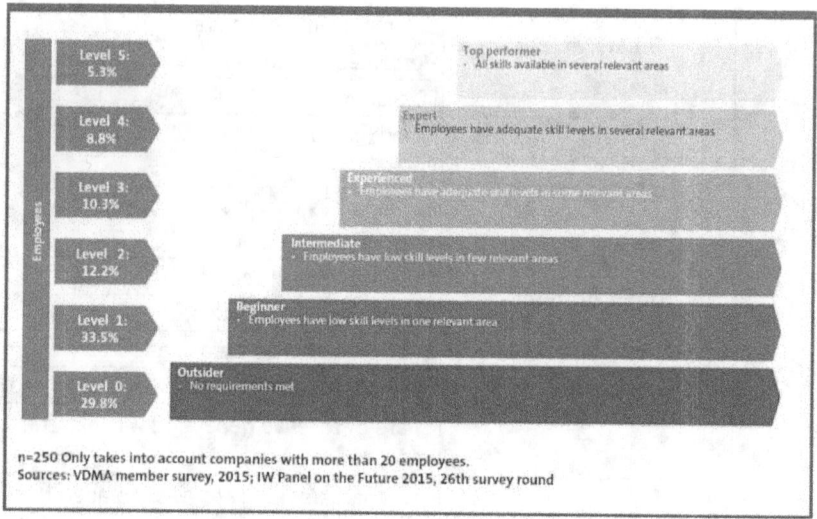

Figure 8-1: Readiness levels in the dimension of employees

Source:
https://industrie40.vdma.org/documents/4214230/26342484/Industrie_40_Readiness_Study_152949800791
8.pdf/0b5fd521-9ee2-2de0-f377-93bdd01ed1c8

The figure above (Figure 8-1) provides actual survey results of the 2015 study as well as detailed criteria for each level in the study's Employees dimension. The percentages of survey respondents that operate at various readiness levels are indicated in the chevrons at the left side of the image.

Table 8-2: Action items for newcomers (readiness 0 & 1)

I4.0 dimensions	Strategy & organization	Smart factory	Smart operations	Smart products	Data-driven services	Employees
Current main hurdles (from Readiness Model)	■ Industrie 4.0 plays little or no role in the strategic orientation	■ Equipment infrastructure not connected to higher-level IT systems ■ Machine and process data not collected	■ Little or no in-company, system-integrated information sharing	■ Products feature little or no ICT add-on functionalities	■ Focus on traditional products; data-driven services either not offered or not integrated with customers	■ Little or no industrie 4.0-specific employee skill sets
Obstacles	■ Uncertainty about economic benefit ■ General uncertainty about I4.0			■ No market need ■ Lack of skills		
Action items	■ Create awareness for I4.0 ■ Define strategies	■ Gradually connect equipment infrastructure to higher-level IT systems	■ Plan initial steps for in-company and external system-integrated information sharing	■ Analyze potential of ICT add-on functionalities	■ Realign product portfolio ■ Define data-driven services	■ Run systematic needs assessment ■ Adapt training and professional development programs

Original layout

Source: htps://industrie40.vdma.org

103

The Industrie 4.0 Readiness study also includes a series of recommended action items (Table 8-2) designed to help manufacturers improve their readiness status. After making detailed readiness measurements, assessment users determine their overall readiness category from these options:

- Newcomers: Levels 0 and 1
- Learners: Level 2
- Leaders: Levels 3-5

The study authors used company demographic information gathered from the survey respondents. The data documented the major obstacles to and provided recommended action items for survey respondents in each of these categories. For example, here are the action items recommended for companies in the newcomers' level of overall readiness:

These are the format, structure, and methods used in the readiness assessment conducted by German mechanical and manufacturing engineers. A group of students and their legal and business partners in the United Kingdom took another approach.

8.4   Warwick University Readiness Assessment

The Warwick Manufacturing Group (WMG) is a department of the University of Warwick in the United Kingdom. In 2017, WMG students and faculty joined forces with global management company, Crimson & Co, and international law firm, Pinsent Masons. The group's Industry 4.0 readiness assessment was a product of this collaboration.

By working with industry and private sector partners, WMG team members added academic discipline to the identification, and possible solution, of a complex business problem — describing the Industry 4.0 readiness of UK manufacturers.

In 2017, the study authors recognised that during the previous 20 years, distinctions between a physical product and a pure service were becoming blurred. Improved IT connectivity, IIoT technologies, and cyber-physical systems create new and innovative opportunities to replace the purchase of a product with a service.

Similar to the Industrie 4.0 Readiness study, the WMG assessment process reviewed the capabilities that enable manufacturers to generate value via Industry 4.0 technologies, ideas, and methods. Like the German study, the UK authors used gap analysis and an Industry 4.0 maturity framework to help businesses identify their progress and needed improvements. In both cases, a survey-based study provided the data needed to build a self-assessment tool.

Major concepts and their relationships — Unlike the Industrie 4.0 readiness study, the WMG team created a model that emphasised supply chain strategy as a driver of business transformation.

This approach led to a framework of six dimensions and 34 sub-dimensions, which the study calls fields. Table 8-3 presents the framework and its topics in tabular form.

Table 8-3: University of Warwick Readiness Assessment

University of Warwick Readiness Assessment	
Products and services	Product customisation
	Digital product features
	Data-driven services
	Product data use
	Data-driven services share of revenue
Manufacturing and operations	Technology (M2M) integration
	Autonomous workplace
	Operations data collection
	Operations data use
	Data modelling/I4 equipment readiness
	Self-optimising processes
	Digital modelling
	Cloud solution use
	IT and data security
Strategy and organisation	Degree of strategy implementation
	Performance measurement
	Investments
	Workforce skills and knowledge
	Collaboration
	Leadership
	Finances
Supply chain	Inventory control
	Supply chain integration
	Supply chain visibility
	Supply chain flexibility
	Lead times
Business models	As-a-service models
	Data-driven decisions
	Real-time tracking
	Real-time and automated scheduling
Legal considerations	Contracting methods
	Risk management
	Data protection
	Intellectual property

The WMG and Industrie 4.0 readiness studies use the same general approach: start by assigning important functional areas (dimensions) such as supply chains. Next, assign each dimension with important capabilities such as supply chain flexibility and inventory control. Finally, use the capabilities to build a graded system of indicators, such as those provided for the Products and Services dimension in the table shown above.

When assessors measure their readiness in each of the six dimensions and 34 indicators, they create a detailed, numerical record of their Industry 4.0 status. Templates and benchmarking enable users to compare their company's findings with those of their industry competitors. Figure 8.2 below, is an example taken from the Products and Services section of the study.

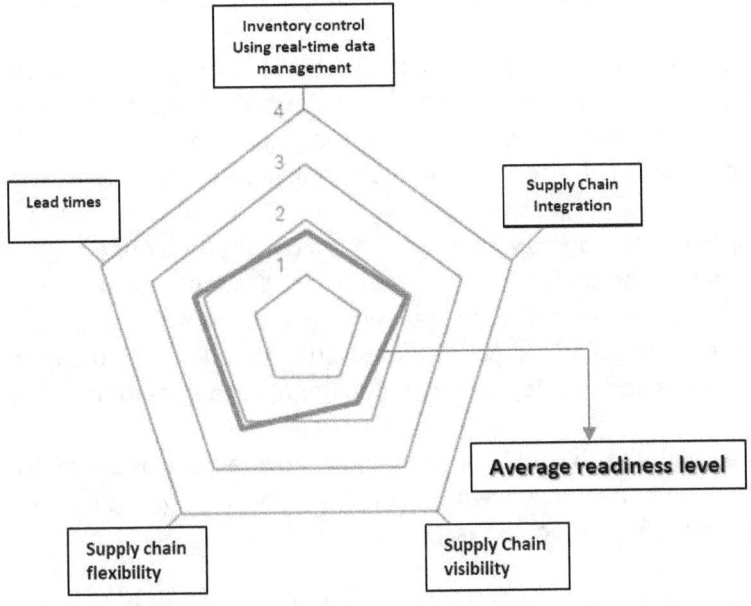

Figure 8-2: Average readiness level

Source:https://warwick.ac.uk/fac/sci/wmg/research/scip/reports/final_version_of_i4_report_for_use_on_websites.pdf

The German and UK assessment tools come from Europe, the original home of Industry 4.0. However, an equally comprehensive and much more contemporary readiness assessment tool has been produced in Singapore.

The detailed review of its content and approach in Section 8.5 is worthwhile because it reflects attitudes and challenges that are much closer to home for ASEAN manufacturers.

8.5 Singapore Smart Industry Readiness Index

Launched in November 2017, the Singapore Smart Industry Readiness Index (SSIRI) is the result of collaboration between national government agencies and members of the private sector. The country's Economic Development Board appointed TÜV SÜD, an inspection, certification, and training company, to work with them to develop a comprehensive but easy-to-use self-assessment tool.

The SSIRI tool is designed especially for manufacturing companies in Singapore. From the beginning of the index's development, the goal was to design, build, test, and share a comprehensive tool for companies of all sizes and manufacturing industries.

Together, the agency specialists and consultants created a working model of the SSIRI. They also selected 300 multinational companies and SMEs in Singapore for training and assistance in using the index. The goal: to help Singapore's manufacturers understand their Industry 4.0 status and develop a roadmap to improve their readiness.

The results of this 300-company pilot provide benchmark metrics for other manufacturers, who want to improve their operations in different assessment areas.

Source materials and major ideas — Index source materials included industry reports, landscape studies, business surveys, and models produced by leading European professional associations and players in the Singaporean manufacturing industry. The index draws on the Reference Architectural Model for Industry 4.0 (RAMI 4.0) developed by Platform Industrie 4.0, one of the world's largest Industry 4.0 networks. RAMI 4.0 is acknowledged by key experts and respected associations to be the reference architecture model that best embodies key Industry 4.0 concepts.

Other reference materials included but were not limited to the Industrie 4.0 Maturity Index developed by the German Academy of

Science and Engineering (Acatech) and the Bersin model for human capital development, developed by Deloitte.

Using the SSIRI — SSIRI authors summarise the high-level process of using the index with the acronym, **L-E-A-D**. Each letter indicates a process step.

"Learning" key Industry 4.0 concepts — The index aims to help companies strengthen their institutional knowledge about Industry 4.0. So, to start, prospective Industry 4.0 adopters should learn and understand key Industry 4.0 concepts. The goal: to build a common understanding and language, which extends throughout and between manufacturing businesses.

Next, adopters should become familiar with the basic Industry 4.0 vocabulary. Measurement of Industry 4.0 readiness requires familiarity with the building blocks, pillars, and dimensions that populate the tool's framework.

"Evaluating" current readiness — The SSIRI provides several guidelines, which were designed to make the evaluation process a bit easier.

- **Scope and timing** — Companies need to define the scope of their assessment and the SSIRI tool can accommodate evaluation of an entire manufacturing facility all at once. Adopters can also use their first forays into Industry 4.0 adoption as a pilot, in a phased approach that enables users to review, equip, and test a smaller unit of their factory or business — one product group, at a time, for example.
- **Stakeholders** — Due to the comprehensive nature of the index, stakeholders should ideally come from a cross-functional team, including key stakeholders such as the plant general manager.
- **Judgment calls** — Readiness measurements in the SSIRI use numerical data, but qualitative judgement comes into play,

too. Sometimes, two levels of a company's readiness status overlap. When this occurs, it's the company's decision which level to assign an indicator.

The index also encourages users to keep these principles in mind throughout the evaluation process:

- **Assessments are snapshots** — Assessment results provide evidence of a facility's current state. They do not infer a facility's potential to generate value in the future.
- **New concepts and ideas** — the index is built around Industry 4.0 concepts as they were understood when the tool launched in late 2017. In the future, readiness assessments should include new manufacturing concepts and technologies as they appear in the marketplace.
- **Please, no cherry-picking** — Assessors should measure all aspects of manufacturing described in the index. Cherry-picking, the practice of measuring some indicators but not others will skew the results and make the index less effective. The index appreciates that the relevance of each pillar and dimension will vary for each company — both today and in the future.
- **No need to maximise** — Assessors are discouraged from using the index to achieve level 5 readiness for all indicators in all dimensions. Companies will benefit more if they select and work towards a higher level, based on their specific business goals and requirements.

Readiness measurements are just part of a broader process of creating a long-term Industry 4.0 strategy.

"Architecting" an end-to-end transformation strategy — The third step is to build a comprehensive transformation strategy and implementation roadmap. Companies can use the index in two ways.

- First, the index provides a checklist, which can help companies ensure that they measure readiness for all relevant aspects of Industry 4.0 adoption.
- SSIRI also doubles as a step-by-step improvement guide, which identifies the intermediate steps needed to adopt Industry 4.0 technologies and practices. Frameworks that describe Industry 4.0 ideas and goals exist, but few provide practical information about how to engage in the Industry 4.0 transformation process.

The SSIRI addresses this deficiency by providing explicit definitions and descriptions for all relevant Industry 4.0 concepts, indicators, and manufacturing and business capabilities. It enables companies to identify their highest-impact initiatives and build effective implementation plans with clearly defined goals and timelines.

"Deliver" and support transformation initiatives — After a company has come up with its transformation roadmap, the next step is to add various infrastructure, cyber-physical system, and process information.

The SSIRI serves as a guide to companies measuring, adapting, and refining their long-term Industry 4.0 readiness statistics. Each company will determine its optimal approach to achieving readiness outcomes with its own implementation steps and initiatives. As mentioned earlier, companies can scale and time their Industry 4.0 improvements according to their business goals and constraints. A one-step, Big Bang implementation is not required.

What's in the Smart Industry Readiness Index? — Like the other Industry 4.0 assessment frameworks described so far, the SSIRI is a self-assessment tool. Its structure and content resemble the German and British studies.

The key difference of the SSIRI, however, lies in how closely its major concepts and dimensions resemble those of original Industry 4.0 source materials. Figure: 8.3 illustrate the relationships of Industry 4.0 concepts as they are used in the SSIRI.

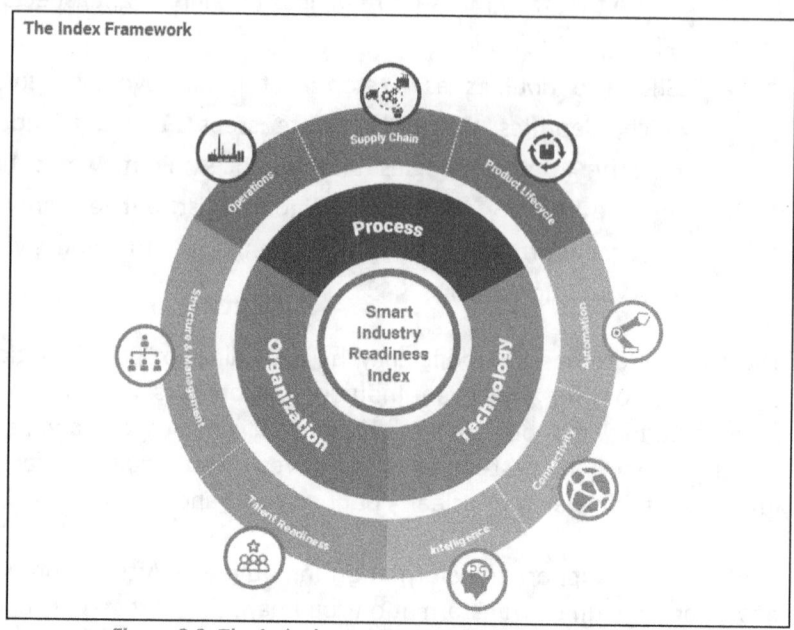

Figure 8-3: The index's concepts are its building blocks

Source: https://www.gov.sg/~/sgpcmedia/media_releases/edb/press_release/P-20171113-1/attachment/The%20Singapore%20Smart%20Industry%20Readiness%20Index%20-%20Whitepaper_final.pdf

8.5.1 SIRI building blocks

The Process building block — Technology adoption cannot occur without effective, well-designed processes. The goal of Industry 4.0 technology adoption is to reduce costs and time to market by connecting intelligent facilities with every part of the production value chain. Opportunities for process improvements occur throughout:

- **Manufacturing operations** — Using Industry 4.0 technologies and practices enables manufacturers to achieve their long-

112

time goal: to create the best quality goods possible for the lowest possible cost.

- **The supply chain** — In Industry 4.0 environments, it's possible to improve manufacturing and business processes from raw material points of origin to the ultimate consumer. This expanded range of processes benefits all players throughout the value chain.
- **The product life cycle** — it's possible to make processes more efficient in each of the stages that every product goes through, from the earliest design stage to manufacture, use, and eventual removal from the market.

From product design studios to the factory floor and many stages of logistics, the wide-ranging changes that occur in Industry 4.0 environments require a high degree of automation, intelligence, and connectivity. That's where new technologies come into play.

The Technology building blocks — New technologies—and new use cases for familiar ones—are the essential elements in transforming manufacturing processes. Manufacturers rely on technology to provide three basic capabilities throughout their Industry 4.0 systems. These include:

- **Connectivity** — measures the state of connectedness between equipment, machines, cyber-physical systems, and humans. As each year passes, more wired analogue devices are being converted to wireless, digital ones. It's the digital, internet-enabled devices that are put to work in smart factories and their offices.
- **Intelligence** — the ability of advanced data management technology to enable high-speed, high-volume data analysis. Intelligence enables responsive, data-driven decision making, the foundation of efficient manufacturing operations.

113

- **Automation** — the ability of technology to monitor, controls, and execute the production processes and delivery of products and services, with little or no human intervention. Intelligent automation, a hallmark of Industry 4.0, enables swift responses to change in technical and business environments.

Including humans in systems that consist primarily of technology and all manner of hardware is characteristic of Industry 4.0. Human organisations and management practices share a vital role in digital transformation.

The organisation building block — It's possible to measure Industry 4.0 readiness in terms of how highly developed organisational and management practices in a manufacturing company might be.

Organisational criteria include:
- **Talent Readiness** — Developing the skills and knowledge needed to operate smart factories and their supporting business processes has always challenged Industry 4.0 adopters.
- **Structure and management** — Structural criteria measure whether a company has a system of rules and policies that guide how roles and responsibilities are assigned, controlled, and coordinated.

Manufacturers have several choices to develop required skills. Formal education and training, internal training programs, or ad hoc on-the-job training. Whichever methods a company might adopt, success depends largely on a company-wide commitment to continuous training and self-education.

Companies wanting to assess their Industry 4.0 readiness apply the structure and concepts of the SSIRI framework to their own factory operations and business processes. Measurement occurs in a series of

114

16 dimensions. This third-level information describes and assigns maturity levels to information and operations technologies, processes, and organisational practices.

SSIRI assessment dimensions — Far from any grand vision of Industry 4.0, digital transformation occurs with the use of and experience with specific processes, technology use cases. Many ASEAN manufacturers, especially SMEs, have limited resources and must be cautious in how they invest their Industry 4.0 development. Unveiled at the Hannover Messe in 2019, the matrix was developed with knowledge partners McKinsey & Company, SAP, Siemens, and TÜV SÜD.

To measure Industry 4.0 readiness, look for it in offices, on the factory floor, and beyond, throughout product life cycles and supply chains. Here's a graphical (Table 8-4) representation of SSIRI structure and content.

Many Industry 4.0 concepts appear in the framework. In addition to the building blocks and pillars mentioned earlier, notice that vertical, horizontal and product integration —important elements of RAMI 4.0— appear in the Process building block. Smart factory automation, connectivity, and intelligence capabilities operate at the shop floor, enterprise, and facility levels of operation. While collaboration, an essential part of Industry 4.0 success, is measured within and beyond individual companies.

Measuring Readiness in SSIRI — As in the German and British readiness tools, the SSIRI provides a table for each of its dimensions. Each table provides detailed descriptions of different levels of experience with Industry 4.0 processes, technologies, and organisational practices. Here's the Supply Chain Pillar - horizontal integration dimension, taken from the Process building block (Table: 8-5). Assessors go through each of the 16 dimension-level tables and choose the maturity level that best applies to their current operations.

Table 8-4: SSIRI 16 Dimension of Assessment

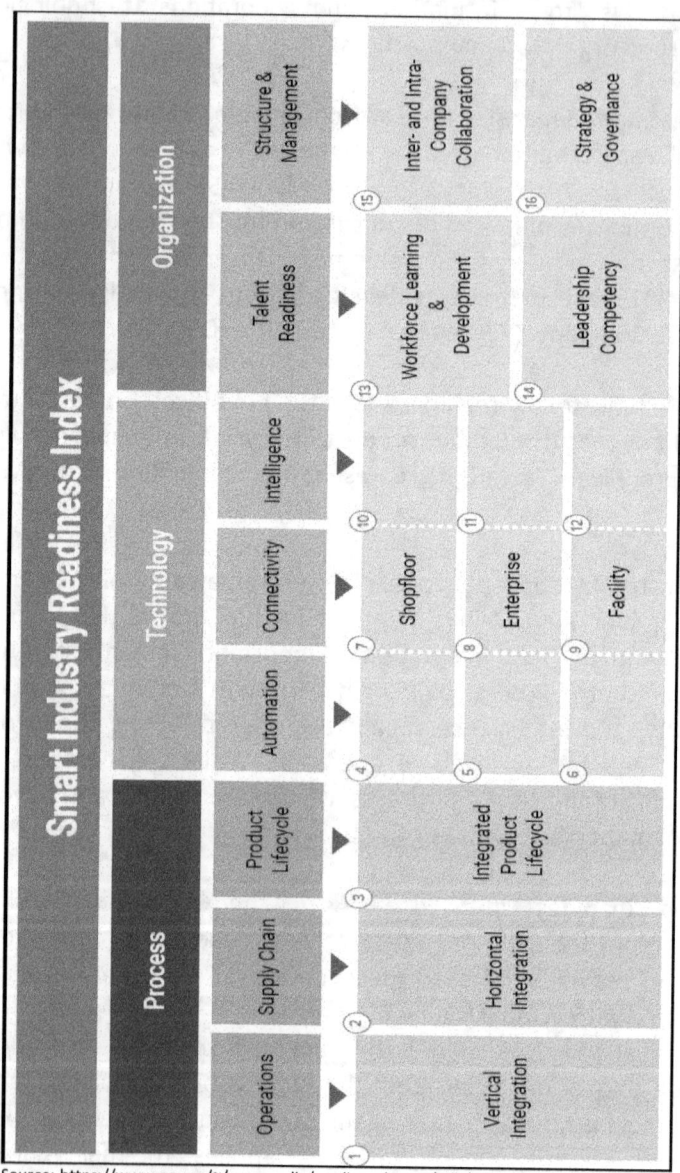

Source: https://www.gov.sg/~/sgpcmedia/media_releases/edb/press_release

Table 8-5: Horizontal Integration readiness in SSIRI

Process Building Block | Supply Chain Pillar | Horizontal Integration Dimension			
Horizontal Integration is the integration of enterprise processes across the organization and with stakeholders along the value chain.			
	Band	Definition	Description
0	Undefined	Supply chain processes[3] are not explicitly defined.	Enterprise processes are managed and executed in silos, based on informal or ad-hoc methods.
1	Defined	Supply chain processes are defined and executed by humans, with the support of analogue tools.	Enterprise processes are managed and executed in silos, based on a set of formally defined instructions.
2	Digital	Defined supply chain processes are completed by humans with the support of digital tools.	Enterprise processes are managed and executed in silos by IT systems.
3	Integrated	Digitized supply chain processes and systems are securely integrated across business partners and clients along the value chain.	IT systems managing enterprise processes are formally linked; however the exchange of data and information across different functions is predominantly managed by humans.
4	Automated	Integrated supply-chain processes and systems are automated, with limited human intervention.	IT systems managing enterprise processes are formally linked, with the exchange of data and information across different functions being predominantly executed by computer-based systems.
5	Intelligent	Automated supply chain processes and systems are actively analysing and reacting to data.	IT systems are integrated from end to end, with processes being optimized through insights generated from analysis of data.

Source: https://www.gov.sg/~/sgpcmedia/media_releases/edb/press_release/P-20171113-1

However, this isn't the end of the Industry 4.0 readiness story. There's another tool, developed by popular demand—Singapore's Smart Industry Readiness Prioritisation Matrix.

117

8.5.2 SSIRI Prioritisation Matrix

April 2019, Singapore's Economic Development Board released a second readiness-related tool, the Readiness Index Prioritisation Matrix. As its name implies, the tool helps manufacturers worldwide focus on and prioritise specific areas of their own digital manufacturing initiatives.

The matrix borrows from the concept of IT system maturity, a familiar analytical tool for almost 15 years. The prioritisation process uses facility-level Industry 4.0 maturity data and current financial performance information.

The maturity approach helps matrix users close the gap between knowing about Industry 4.0 solutions and knowing how to use its basic principles and methods in their own smart facilities. The idea is to use the matrix as a guide to creating more systematic, thorough, and consistent initiatives.

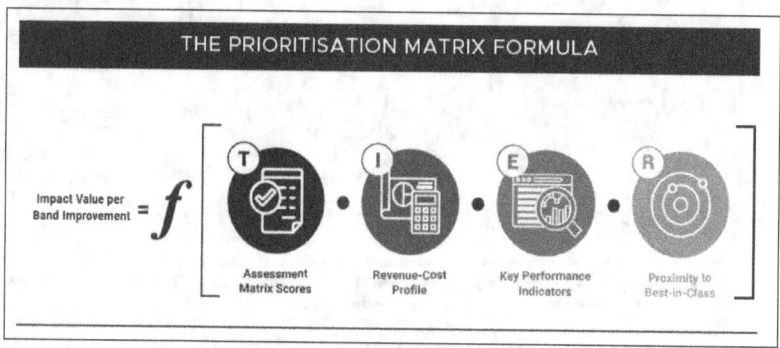

Figure 8-4: The Prioritisation Matrix Formula

Source: https://www.gov.sg/~/sgpcmedia/media_releases/edb/press_release/P-20171113-
1/attachment/The%20Singapore%20Smart%20Industry%20Readiness%20Index%20-%20Whitepaper_final.pdf

How the prioritisation matrix works — The prioritisation matrix uses four sets of metrics to calculate a numerical score. The number represents the relative benefit that a company will gain when digital transformation initiatives focus on a specific business or technical area.

This approach helps manufacturers get a clear picture of the feasibility of their desired outcomes. It can also make developers more confident that the projects described in the tool could generate enough value to satisfy smart factory business goals.

The matrix uses the following types of information to help manufacturers judge the priorities of digital manufacturing projects and initiatives:

- **Assessment matrix scores** — These metrics reflect a manufacturer's Industry 4.0 maturity. Values range from Band 0 to Band 5, across all 16 dimensions
- **Revenue-cost profiles** — These numbers reflect company cost category line items as a percentage of total revenue.
- **Key performance indicators** — KPIs measure a company's ability to achieve its business and strategy goals.
- **Proximity to industry best-in-class** — The assessment matrix assigns a maturity level (band number) to a company's capabilities. This category measures the distance of a company's maturity to that of best-in-class operators.

Manufacturers use these metrics in a formula, which provides an impact value per band improvement (the impact value) for each of the 16 assessment index dimensions. Each Impact value reflects the relative benefits that a company gains improving its maturity by a single level (band), within a specific assessment index dimension. By comparing impact values of the different maturity improvements, manufacturers can use quantitative data to identify the high-priority areas to develop.

9 Technology Adoption in ASEAN

"The adoption of new technology finally occurs when ease of use, economic savings, and trust all come together to work toward change."

EVERYTHING — Alec J. Ross

When we discuss Industry 4.0, sooner or later, we're talking about technology. The topic might be buying new equipment or upgrading legacy machines with Industrial Internet of Things-enabled devices. Whatever the topic, Industry 4.0 involves putting technologies to work throughout product life cycles and the supply and value chains.

Each of the nine technologies that we describe here presents a different profile of use cases, capabilities, and power to generate value. Each profile includes the technology's manufacturing-related functions as they appear throughout the RAMI 4.0 model. Finally, we highlight some early uses in manufacturing processes, such as design, R&D, production, and post-manufacturing tasks. So, which technologies are ASEAN-6 nations using in their quest for digital transformation, productivity, and profits?

Slowly but surely, shop floors and supply chains throughout ASEAN are becoming ground-zero for innovative manufacturing and logistics processes. Here are examples of how Industry 4.0 technologies work and what they can do for manufacturing companies of all sizes.

9.1 Additive Manufacturing (3D Printing)

In increasingly customised processes, 3D printing methods are creating tools, fixtures, and jigs onsite, in less time than with mass-production methods and without the need to outsource fabrication services. This approach is one example of how forward-looking manufacturers make unglamorous objects part of their path to digitalised production.

So, what is Additive Manufacturing? Its alternative name, 3D printing, tells the story. Traditional manufacturing is a subtractive process. Making products or components requires cutting, forming, polishing, and otherwise altering raw materials by reducing their volume and wasting the bits left behind.

3D printing, however, is a very precise additive process. Raw materials are applied in layers into whatever form the design requires. There's no cutting, leftovers, or need to recycle waste materials. The additive production process reduces waste, processing time, and equipment-related resources.

9.1.1 Making Inroads in the Manufacturing Process

3D printing has opened new opportunities for more efficient production, factory maintenance, and R&D. Take R&D, for example.

When the average person imagines "research and development," they think of numerous traditional prototyping methods, often utilising handcrafted or moulded models. It's an approach that uses a lot of design time and money. Using 3D printed prototypes saves design costs and provides physical models that stakeholders can touch and examine closely. They tend to be much more persuasive than the best computer rendering.

What's more, being able to make prototypes easily and cheaply enables companies to get creative. Lower design costs might give the budget extra funds for a bit more trial and error developing new designs.

9.1.2 Taking the Pain Put of Parts Replacement

What about factory maintenance? Think of the nightmare known as parts replacement. OEMs discontinue replacement parts after a pre-set number of years. If a part's availability time runs out, factory procurement specialists must turn to injection moulding or metal stamping companies, which is usually an expensive process. 3D printed parts can provide lower-cost replacements and eliminate the delays and pain of our nightmare scenario.

First, say good-bye to minimum orders. Most 3D printing companies don't have minimums because they can sell less than a dozen parts and still make a profit. Second, a 3D printing specialist can digitally reverse engineer and strengthen the part. So, say goodbye to the high costs of prototypes, too.

Finally, 3D printers can make new parts in a few days (sometimes less). Meaning manufacturers, who do their predictive and preventive maintenance chores, can get replacement parts without losing tens of thousands of dollars per minute to a production line breakdown.

9.1.3 3D Printing-Tech Development

3D printing will probably take over some R&D and maintenance tasks soon. Don't expect the process to eliminate all production processes though. Many traditional manufacturing methods are too complex, operate on too large a scale, or require materials that 3D printing can't use.

On a broader note, one might question whether current 3D printing equipment qualifies as Industry 4.0 technology. After all, most 3D printers are standalone equipment, which provides few, if any interfaces to larger manufacturing ecosystems.

To become truly Industry 4.0-ready, 3D printers must connect to control and operations devices. When equipped with the required interfaces, a fully automated 3D printer could operate entirely via software, without human attention.

Looking forward to future 3D printing capabilities, it's likely that the technology will:

- Decrease the time to market of new products by reducing prototyping cycle times.
- Enable more product customisation because some 3D-printed tools were unavailable with traditional tooling methods.
- Provide higher-value-added products at an affordable cost.
- Enable businesses to confirm designs before committing them to production. This ability would help manufacturers avoid the risks of wasted materials, human error, and uneven product quality.
- Make supply chains more efficient by producing components onsite to reduce wait-time for parts. The ability to make some parts locally would help manufacturers avoid lengthy delivery times and respond more quickly to changing customer preferences.

9.1.4 3D Printing Adoption Trends

ASEAN manufacturers view 3D printing as an on-demand manufacturing tool. It provides fast, flexible production and rapid response to unpredictable changes in demand.

Four 3D printing service providers offer ASEAN-6 manufacturers a virtual inventory of parts and on-demand production capabilities.

- Fast Radius provides an early-stage example of Industry 4.0 business model innovation. In late 2016, the California-based company opened a factory at a United Parcel Service site in Singapore. The company uses its Fast Radius On-Demand Production Platform to produce industrial parts, which they deliver to manufacturers through the UPS transport network.

- The company's service enables customers to virtualise their inventory and avoid making and storing parts for an unknown future date. Instead, customers can produce a smaller number of parts and reduce lead times. The entire process is fast and on-demand.
- 3DHUBS, a Thai additive manufacturer, offers rapid turnaround 3D printing, CNC machining, and injection moulding services for the ASEAN market. The company works with a network of local 3D printing companies, supervising and curating their vendors' output.
- Thaisakol Group Additive Manufacturers, another Thai company, has entered the additive manufacturing market. They provide 3D printing services for metal parts.

Additive manufacturing is getting a foothold in ASEAN nations, but it's still in a very early stage. The proof of concept will occur in the future, when more manufacturers order more additive manufacturing equipment to make their own parts.

9.2 Augmented & Virtual Reality (AR/VR)

The development of AR and VR technologies has been driven by widely available, low-cost 3D graphics hardware and advances in high-bandwidth internet technology. Their use in manufacturing enables companies to improve workforce skills, ensure more consistent product quality, reduce design and production costs, and slash the time needed to develop a product from concept to production and delivery.

ASEAN manufacturers are just beginning to kick the virtual (reality) tires. Early signs of lower error rates, better productivity, and remote collaboration between humans and machines are promising signs of future performance. However, it's going to take many success stories and persuasive efficiency metrics to usher in a time of strong adoption.

Manufacturers, who must justify the investment of AR/VR with robust business cases, are investigating the potential impact of AR in their production processes. A regional survey indicates that they should count on using 50 per cent or more of their investment in the deployment of existing resources, mostly in the form of people's time.

There's evidence, however, that taking the plunge on AR and VR technology yields strong process improvements. Some companies deploying AR glasses in pilot projects have reported up to an 80% reduction in human error and 20-to-30-per cent returns on investment.

9.2.1 Virtual Reality Capabilities in Manufacturing

In the manufacturing sector, VR is a training simulator that enables users to experience a totally virtual world. Users experience the virtual environment as if it were real, making VR a 3D experience that can be shared with other people and computers.

VR enables users to explore and focus on virtual objects at levels of detail that are appropriate to their job on the factory floor. VR makes it easier to learn work and production processes by presenting information in a simulated, three-dimensional space. VR is an especially good choice as a training aid, especially when:

- Trainers must instruct many employees distributed over a large geographical area.
- All employees in a group of trainees must get the same instruction.
- Specific skills are best learned by experiencing very realistic conditions.
- Specific skills are best learned when employees are permitted complete access to an entire facility.

- Trainees must practice new and existing skills in a safe, simulated environment.
- Trainees must see how a product takes shape as it moves through manufacturing systems.
- It's too complicated, dangerous, or expensive to train employees in real settings.

9.2.2 Augmented Reality's Important Role

Unlike VR's fully-immersive digital reality, augmented reality blurs the line between what's real and what's computer-generated. AR can lay holographic images, instructions, or data atop an individual's real-world view of equipment, workstations, warehouse locations, or production lines. This mixed data-and-image approach enhances what trainees see, hear, and feel.

AR applications in manufacturing often combine the capabilities of the IIoT and augmented reality. For example, engineers can view renderings of parts, part numbers, cables, and bolts and instructions on how to assemble a specific component. By using AR methods, trainers can layer existing production location information with an ocean of instructive virtual planning objects.

Collaboration and remote assistance are other useful applications for augmented reality. The technology helps users in different locations to see the same thing at the same time. For example, an expert could show a maintenance technician in another facility how to fix a broken part.

Figure 9-1: Application of AR in manufacturing environment.

Augmented reality enables manufacturers to:

- Train employees for their production line jobs.
- Optimise the flow and location of production lines.
- Position robots, production cells, people, and automation lines to make production more efficient.
- Handle and operate virtual tools and equipment.
- Plan layouts around support pillars, lighting, and heating and air-conditioning ducts.

9.2.3 Value of AR & VR based Training

Augmented and virtual reality training environments offer many important capabilities, such as:

- Providing step-by-step, visual and oral instructions in real-time.
- Helping workers navigate warehouses and factories more efficiently.
- Identifying the right tools and parts needed to complete a task.

- Correcting mistakes and providing real-time feedback during training.
- Overlaying key performance metrics and operational data on a trainee's view of specific pieces of equipment.
- Helping workers complete unfamiliar or challenging tasks or troubleshoot problems with expert guidance.

Augmented and virtual reality-based training can transform the way controls engineers are trained and evaluated. These methods can also enrich the wisdom harvest process that occurs when retiring workers leave the factory staff.

In 2018, interviews with Swiss manufacturers highlight significant technical and human challenges to companies adopting AR and VR applications on the factory floor:

- **Limited budget and human resources** — Many manufacturers cited their lack of expertise developing AR and VR applications as a problem. Others mentioned that workers are already absorbed by their daily tasks and responsibilities, and therefore wouldn't be able to handle the additional effort that comes with the introduction of AR and VR applications.
- **Limited knowledge and experience** — Some companies lack experience and production metrics needed to operate AR applications effectively. Even judging the effort required to build AR and VR applications has proved challenging. Furthermore, some company decision-makers can't quite imagine what these new technologies could do on their shop floor. That makes it difficult for technical decision-makers to build a solid business case and predict a reasonable return on investment for business leader stakeholders.

9.2.4 AR & VR Adoption in ASEAN

Commercial IT and business applications of AR/VR technology are becoming available. The most commonly cited use cases include product design, field maintenance, production process training, and marketing applications.

Across ASEAN, SMEs are starting to explore new technologies to catch up with their bigger rivals. Manufacturers in ASEAN-6 countries are slowly gaining experience with AR and VR, experimenting with how it can improve competitive advantage and productivity in their operations.

The ASEAN market is home to many vibrant start-ups and emerging AR and VR companies. However, few of these companies are manufacturers, and fewer are using AR and VR in production or training processes. Most reports of AR and VR technology remain at early stages of development—new partnership deals and informational workshops, such as the one sponsored by the Malaysia Automotive, Robotics and IoT Institute.

Occasionally, however, global enterprises such as BMW are adopting these technologies to improve production. In early 2019, BMW started to use VR and AR to improve training, production, and workflow planning. Data collected from several years of production operations provided the information needed to build the apps.

9.3 Advanced Robotics

Industrial robots are nothing new. For decades, they've become fixtures of Southeast Asian production lines. Now, however, robots are much more than mechanical-muscle and limited digital intelligence. Manufacturers are engaging more with smart robots, products of the fusion of robotics and varying degrees of artificial intelligence.

Many ASEAN countries are using advanced robotics to maximise efficiency and competitive advantage of their manufacturers. Implementing manufacturing robots in the region can boost manufacturers' competitive edge by enabling lower production costs, shorter production cycles, less hazardous workplaces, and better product quality.

In Southeast Asia, the electronics and automotive industries are the largest users of this advanced robotics. Growth in the region's consumer electronics industry drives the demand for robotics investment, which industry observers estimate will total about $475 million by 2021. Regional electronics industries use robots in many processes, including materials handling, inspection, testing, assembly, packing, and assembly.

The integration of collaborative robots (also known as cobots) into production lines continues to help businesses across ASEAN to improve their manufacturing processes and remain competitive. Unlike traditional industrial robots, which are bulky and fast, cobots are lightweight and mobile, a bit slower than robots, more affordable, and reprogrammable for different applications. In Southeast Asia, cobots work the line in the electronics, automotive, pharmaceuticals, and chemical industries.

9.3.1 Social Impact of Robots

The introduction of more robots might change manufacturers' production strategies. Relocating facilities to continue playing the low-wage card is unlikely to be as profitable as previously. With the advent of heavy automation via robots and cobots, labour costs will have a much lower impact on total manufacturing costs. This change makes it likely that foreign manufacturers would relocate their plants to be closer to their customers. If this change occurs, it could achieve significant time and logistics cost savings—and inflict a body blow to ASEAN manufacturers.

9.3.2 Capabilities of Robotics in Ind 4.0 Settings

Robots already monitor the production line and adapt to changes in the production process. Now, it's possible to network several robots that can do their own work and the work of robot and human colleagues.

Cobots are valuable automation tools, which help increase productivity, improve product quality, and promote employee safety. Thanks to advances in vision and sensor technology, cobots safety intelligence makes it possible for them to collaborate safely with people. Straightforward, user-friendly programming makes it easy for unskilled workers to reprogram their machine colleagues with no data science degree required.

The most exciting development is that robots guided by sophisticated algorithms and complicated instructions, learn what to do directly from the data without relying on a pre-set and fixed equation or program. This is another case in a growing list of "no humans required" examples.

Many other technologies enhance robot capabilities. Advanced cobots may use machine vision, AI, natural language processing, 5G communications, mixed reality, LiDAR sensors, or many more.

9.3.3   New Support Technologies

Process control used to be limited to programmable logic controllers (PLCs). Their complexity and rigidity made them costly to set up and reprogram when process design changes were needed. No longer. Behaviour tree technology smoothly blends logical flows and actions that command robots to perform tasks. It's all done with on-board software, controlled by a human operator, who knows the best workflow for each task.

Starting with behaviour-based robots, manufacturers can build an intelligent production line, then several lines, and finally create an intelligent factory. These machines would not only perform tasks but collect and analyse productivity, quality, reliability, and cost data for themselves and other robots in a process line. Robots also provide managers and engineers with an understanding of the data so they can act when needed.

9.3.4 Advanced Robotics Use Cases

Of the technologies described here, robotics would be the most familiar to ASEAN manufacturers. However, the pattern of use is typical of other digital manufacturing technologies—very uneven levels of experience from one country to another. Here's a review of a robotics adoption use case in Indonesia, a mid-tier company in the ASEAN robotics maturity scale.

By adding cobots to their production lines, PT JVC Electronics Indonesia (JEIN) has taken the first step in robotics maturity. First, they've reduced the need for their workers to perform repetitive, menial, and dangerous tasks. For example, workers now avoid dust particles and hazardous fumes when they solder and separate cut printed circuit board parts.

JEIN manufactures car audio-visual and navigation devices. Until they had installed their cobots, most of the work was manual. In 2018, the company deployed seven cobots to work more productively and keep the quality of their products more consistent.

Installation was relatively simple. There was no need to make drastic changes to the shop floor layout, and the cobots supported different end-effectors, such as screwdrivers, soldering irons, and grippers. This flexibility enabled the production team to customise the cobots to perform three tasks—soldering, screwing, and pick and place.
When JEIN engineers made their business case to buy the cobots, safety was an important consideration. The cobots' patented safety system enables humans and machines to work together closely, without setting up safety fences.

The changes seem to be paying off. With the move towards automation, JEIN managers plan to reassign their manual workers to work on higher-value processes. If anyone needs proof of success, how about an annual reduction in annual operations costs, valued at more than $80,000?

9.3.5   Robotics Readiness

So, what's the most likely pattern of robot use among ASEAN manufacturers? From a readiness standpoint, Vietnam and Singapore seem to be best prepared for adopting robots today. Workers' technical skills, the ability to program computers, and STEM (Science, Technology, Engineering and Mathematics) knowledge and skills in these countries rate highly among their ASEAN neighbours.

Less developed ASEAN member states such as Laos and Burma (Myanmar) have low internet penetration rates and poor IT infrastructure generally. These deficiencies will delay adoption of automation and robotics for the foreseeable future. Their economies are likely to be disrupted by the robotics revolution, but it will happen slowly.

That puts Indonesia, Malaysia, Thailand, the Philippines, and Cambodia in the crosshairs of economic and social changes. These ASEAN nations are most likely to experience disruption caused by advanced robotics technologies.

9.3.6   A Call for Policy Help

Robotics specialists in and out of ASEAN governments are calling for the creation of policies that help manufacturers establish STEM education and retraining programs to mitigate whatever disruption occurs. This close collaboration between governments and businesses throughout the region will determine their success in overcoming the challenges of robotics-based disruption.

9.4 Artificial Intelligence & Machine Learning

Implementing Industry 4.0 initiatives involves two phases of activity. The first phase involves connecting instrumentation and all other manufacturing assets. The second step is focused on figuring out what to do with all that data generated by connected machines and devices.

Currently, most ASEAN manufacturers adopting Industry 4.0 are in the connection phase. Farther along the digital manufacturing process, AI and machine learning will have a significant role in helping manufacturers use their data.

In the manufacturing case, the challenge is putting huge quantities of mostly unstructured data to work. Every organisation accumulates data that it can use to improve its operations. However, most organisations in ASEAN still do very little with it to grow value for their companies.

Machine learning software collects and analyses this abundant resource to help business and technology leaders make predictions and meet company goals. Sophisticated Machine Learning (ML) algorithms recognise characteristics and patterns needed to make useful generalisations from the detailed data. Each machine learning system has a clearly defined job: recognise patterns and draw conclusions from them.

9.4.1 AI Applications in Smart Factories

Artificial intelligence refines simple ML pattern-search capabilities into actions that are useful and valuable to humans. AI technology is ready for use in ASEAN manufacturing plants, but few ASEAN manufacturing plants are ready for AI. Many ASEAN manufacturers are still in the stage of connecting the assets on the floor in order to collect the data and gain the visibility. They have yet to begin their digital manufacturing journey.

The goal of digital manufacturing is to automate the entire production cycle. In this highly developed smart factory environment, control systems not only collect data, but create predictions, and perform production tasks. In smart factories, machines, interfaces, sensors, and components communicate. Specialised AI software enables plants to use their production data, create predictive models, and define actions that respond to what the pattern data tells them.

For example, machine learning enables manufacturers to take a different approach to current scheduled / preventive maintenance, under which circumstance machine shut-down is inevitable. In predictive maintenance, instead, algorithms used in big data analytics processes can determine the current status of the asset health in operating context and provide anomaly detection, hence predict the next failure of a device, machine, or system. When the software application receives notification that reflects a high deviation from normal operational pattern or rule, the system will trigger impact analysis and automated response, while informing maintenance personnel to act on the target asset way before the machine failure happens. Being called as asset performance management (APM), the system has been stemmed from classic condition based monitoring (CBM), adding machine learning to dramatically enhance the reliability during the manufacturing operations.

9.4.2   Product Design Capabilities

Manufacturers also use artificial intelligence during the product design phase. Designers and engineers can use generative design software to explore many possible ways to arrange and organise a solution. With this information, designers can specify types of materials, production methods, budget limitations, and production time constraints. Product and process designers can use this information to test their solutions without wasting resources on materials and processes that don't work.

9.4.3 Optimising Processes & Responsiveness

Another promising use of AI lies in the area of machine-business connectivity. This application uses AI algorithms to optimise the manufacturing supply chain by helping manufacturers anticipate and respond to changes in the marketplace.

In order to build software that estimates market demand, an algorithm can learn the current trends and analyse demand patterns related to locations, dates, weather events, socioeconomic attributes, and much more. Manufacturers can then use this information to forecast within the inventory, demand and control, and further optimise warehouse control, staffing, and many other variables. In such cases, manufacturers can schedule more accurate logistics and plan quicker shipping activities, whereby reduce the cost on transportation and shorten the lead time to gain competitive edge.

9.4.4 AI Use in ASEAN Countries

In April 2019, Microsoft Corporation and IDC released Future Ready Business: Assessing Asia Pacific's Growth with AI. The study documents the increasing use of AI among manufacturers in ASEAN and other Asia-Pacific nations.

The study notes that manufacturers' use of AI in profiled countries lag behind that in national economies generally. Their recommendations include strengthening AI-related strategies and data, as well as company cultures that support more innovative, independent thinking.

Nevertheless, AI adoption is still very new and infrequent in ASEAN-6 countries. Its adoption is described as "slow and patchy," but some manufacturers are testing the waters of AI development. Below we discuss some examples.

In Singapore, American computer maker, Hewlett Packard, and the Nanyang Technological University announced the startup of an $84 million digital manufacturing laboratory. Opened in October 2018, the HP-NTU Digital Manufacturing Corporate Lab reflects a partnership between the university and the global computer maker. To start, the laboratory staff and HP researchers will work on 15 projects that use AI/machine learning, cybersecurity, and new materials-related technologies.

Another recent business-education partnership throws light on one of the routes to AI adoption in Indonesia. Owners of the Indonesia-based Bukalapak e-commerce platform and Indonesia's Bandung Technology Institute launched Indonesia's first AI and cloud computing innovation centre in February 2019.

Located at the Bandung Institute of Technology in West Java, the centre will attract students, researchers and lecturers to participate in its projects. Bukalapak also wants to establish AI development hubs in several regions of the country, including Yogyakarta, Medan, and Surabaya. The company is also trying to attract software developers throughout the nation to the institute.

9.5 Blockchain

Born from a deep distrust of banks and financial third parties in 2008, Blockchain technology spawned a cryptocurrency boom in 2017, which then plummeted in 2018 and been fluctuating in 2019. That spectacular decline does not erase the inherent value that Blockchain technology presents to manufacturers.

No one is sure of the influences that Blockchain technology might have on the regional manufacturing sector. Many non-cryptocurrency applications are still being developed, but throughout ASEAN, there are wide gaps in the capabilities and readiness needed to adopt Blockchain technology. So, it's best to consider any reporting of Blockchain use cases as status reports of its adoption throughout the region.

A Blockchain is a digitally distributed ledger — a series of encrypted databases, which are copied and distributed to many computers. Processed data is collected, encrypted, and formed into packages called blocks. A series of these blocks become a chain. Given a block's structure and sophisticated encryption, once the data is in a block, it becomes extremely difficult to add to, delete, or change transaction data without approval from all nodes of the network.

9.5.1 Hackable Blockchain

Very strong resistance to unapproved-editing makes Blockchain a system that's secure but still fallible. Blockchain technology can store, protect, and distribute digital assets wherever there's an internet connection. In early 2019, news of a successful Blockchain "hack" led to a long and convoluted online discussion of what a Blockchain hack is and whether one actually happened.

By July 2019, the verdict of most financial and cybersecurity analysts agreed that Blockchains are getting hacked, in both cryptocurrency and smart contract use cases. However, the general response has been to call for more effective smart contract coding practices, not a wholesale withdrawal for the technology generally.

9.5.2 Platforms for Manufacturing

Although Bitcoin Core, the original cryptocurrency, is the best-known Blockchain application, it's hardly the only one. In July 2019, there were 1,910 cryptocurrencies and any number of other Blockchains in existence. The critical point, however, is that it's possible to use sophisticated encryption and usability to secure and manage business transaction data.

Currently, Ethereum is the most highly regarded Blockchain technology for B2B applications. It's appealing to larger organisations, which take part in networks and exchange sensitive information between several parties. When security is the priority, the right distributed ledger application can provide secure data exchange in a highly scalable system.

Like all technologies, Blockchain technology has changed in the past few years. What's happened?

9.5.3 Blockchain Technology Improvements

Only nine years after the introduction of Bitcoin in 2009, Blockchain tech is already in its 3.0 stage. Stage 1.0 involved passive but very secure data storage. When Blockchain 2.0 emerged, users could embed business logic in the chain itself. Developers embedded the logic in a distributed app platform (DAP), which some people call a smart contract. These smart contracts have pre-set rules and logic, which are triggered by specific conditions.

Programmers developed Blockchain 3.0 practices when they discovered that Blockchain technology couldn't always assure that transactions were private, unchangeable, unambiguous, and scalable. B3.0 is what we call this period of improved capabilities.

What's happening in the Blockchain software market? — First, we're not talking about cryptocurrency markets. Our more general discussion involves some use cases familiar to cryptocurrency buyers. However, many potentially useful Blockchain applications have more to do with process security and control as in smart contracts than with crypto funds.

To say that the Blockchain software applications market is "fluid" is a monstrous understatement. Start-ups and established companies are constantly testing promising ideas and use cases. Blockchain technology development is at an early stage. Business and technical decision-makers should expect a challenge working out which Blockchain products to buy and whether they need it.

A flood of specialised solutions that address specific manufacturing requirements is on its way, however. Getting ready for this volatile environment requires selecting flexible, vendor-neutral platforms and integration capabilities that are quickly deployed, agile, and easy to retire.

9.5.4 Software Giants Make their Bets

Cloud solution providers are making significant investments in cloud-based Blockchain infrastructure. Blockchain-as-a-service (BaaS) in the cloud is now a reality. For example, you can now set up distributed Blockchain infrastructure in the Microsoft Azure cloud environment. Also, Microsoft and other enterprise software giants are adding Blockchain technologies to their core ERP and supply chain products.

Vendors that make middleware, analytics, and system integration software are also building frameworks that integrate Blockchain technologies with their existing line-of-business applications. These new programs use Blockchain data to support real-time processing and event management.

Looking ahead, business decision-makers in ASEAN might want to watch for emerging Blockchain-enabled consortia in their local manufacturing industries. These organisations might emerge in areas such as logistics management as well as manufacturing. There will almost certainly be many Blockchains, which must be orchestrated and added to existing processes to create value.

9.5.5 Using BlockchainTech in Manufacturing

It's important to use Blockchain technology appropriately. The technical overhead and delays associated with writing to a Blockchain makes it unsuitable for manufacturing applications, such as process control. However, using a Blockchain application to record product movements monitored by IoT sensors in real-time is very feasible. Best-bet Blockchain use cases include:

- **Proving ownership** — This is an appropriate use if any user needs to verify ownership of specific assets, such as physical assets, money, or intellectual property.
- **Tracking shipments** — One can combine Blockchain technology with business apps or AI or use signals from IIoT devices to distribute secure transaction data. Instead of possibly losing data on paper, one can share product location and origin data very quickly and securely by using the Blockchain.
- **Proving the origin of assets** — Blockchain technology makes it possible to track components through the supply chain from pre-assembly fabrication to assembled items.
- **Tracing manufacturing and business data** — Tracing components, modules, systems, and products through the Blockchain is a crucial way to reduce business risk.

Transferring location data via the Blockchain helps avoid losing track of finished goods. Blockchain practices also provide plenty of evidence to prove compliance with current laws and regulations.

Blockchain capabilities are expected to appear in process transformation, track and trace, supply chain tracking, warranty management, asset sharing, and other tasks. The major consideration, however, will be whether manufacturers think the value of smart contracts is worth the risk of hacks or breaches.

When it comes to using Blockchain technology, ASEAN manufacturers are still in data collection mode. Whatever news there is comes in the form of forecasts. The tone of articles is generally positive but always framed in the future tense.

9.5.6   The Industrial Internet of Things

Have you ever thought about how the Internet of Things (IoT) transformed itself and found its way onto the manufacturing shop floor? Here's a brief history.

Start with your basic Internet capabilities circa 1995. Integrate many operating systems, standards, and communications technologies, wireless and wired. Before long, it's 2010, and internet-connected RFID devices have morphed into the Internet of Things (IoT). Next, add millions of inexpensive sensors as well as cloud-based data storage and management services. For a final touch, add remote communications protocols.

Voila! Welcome to the Industrial Internet of Things (IIoT). This is the basic framework on which we hang digital transformation technologies such as AR/VR, big data analytics, and advanced security capabilities.

9.5.7   A Global Network of Devices

What we call the IIoT is the global network of connected, physical objects used in industrial facilities. Each object has an IP address and an internet connection. The IIoT enables these objects to communicate with each other and with nonindustrial, internet-enabled devices and systems.

The distinction between the IoTs is simple. The industrial-strength IIoT makes smart factories possible by connecting machines to other machines, devices, data management software, process optimisation apps, productivity software—and humans.

The consumer-centric IoT is something else, consisting of hundreds of millions of devices across broad societal applications. Instead of monitoring the temperature of molten steel, it opens your garage door or helps you watch your baby.

That means that leading manufacturers use IIoT devices to improve the efficiency and productivity of their operations. Connecting older machines or systems to the smart factory IIoT requires retrofitting them to enable sensors. However, the retrofitting process isn't nearly as painful as in the past, and new equipment often comes with built-in IIoT capabilities. Despite much unhelpful hype of IIoT capabilities, it provides value by enabling several rock-solid use cases.

9.5.8 Optimising Processes

Does your process use sensors or monitoring devices on your manufacturing equipment? Does it use numbers to describe that process? If so, it's ripe for optimisation with IIoT equipment and advanced indoor connectivity. In more sophisticated systems, it's also possible to build in intelligence that responds to pre-set process rules if and whenever a change is required.

Companies that must coordinate production throughout several facilities can develop a secure virtual Manufacturing Execution System (MES). This platform provides a carefully planned schedule by using cloud-based, data storage and management services; big data analytics; and IIoT devices. End-to-end component traceability and stored production metrics help the MES predict when each facility should produce a certain number of items.

9.5.9 Developing New Value-Added Services

Hewlett Packard has developed a new service, which solves a long-term customer problem. The company designed a new printer, one that uses an embedded, internet-connected device. The device monitors ink levels autonomously and uses ink consumption data to predict when the printer would run out of it. Device programming enables each printer to order replacement cartridges automatically, just in time, without telling a human.

9.5.10 Testing & Design Simulations

There's nothing like lots of clean, useable data to verify design simulations and models. For example, by using big data analytics via the IIoT, mechanical engineers can ascertain how a machine's vibration sensors behave and where the resonance nodes are. They can compare the machine data with the models and analysis, and builders can create a machine health index, which can help to reduce downtime.

Adding various machines and devices to a factory's IIoT infrastructure requires integrating them with corporate manufacturing resource planning, asset management, or product life cycle management software. Adding an IIoT system to older equipment can soon create the perfect portrait of frustration. This is especially true when older equipment lacks internet connectivity. Good news, however: many sensors that are interoperable with legacy equipment are now available.

A shortage of workers with IIoT-related knowledge and skills is another potential barrier to faster IIoT adoption. Setting up or expanding an IIoT system to accommodate manufacturing processes requires specialised skills. These skills range from IT networking knowledge, device integration capabilities, an understanding of big data analytics, and knowledge about industrial automation. Also, IIoT solutions require processes and data management procedures that many companies don't have yet. You can't build IIoT resources overnight.

9.5.11 IIoT Use in ASEAN Countries

In ASEAN, manufacturers are eager to invest in IIoT devices. To date, it's difficult to find actual IIoT deployment projects in the manufacturing sector. A six-country, 2018 survey by the AIBP indicates that on average, 84 per cent of enterprises in the six leading ASEAN manufacturing nations were exploring or implementing IoT-based solutions.

Singapore's IIoT development is consistent with its long-term goal to become a Smart Nation. The national government plans to invest S$3.2 billion for five years (2016-2020) to support research in advanced manufacturing technologies such as IIoT.

IIoT solution providers active in Singapore include Siemens (Germany), ABB, Schneider Electric, Dell, Emerson Electric, Accenture, OMRON, an automation-solution provider, has opened a $10 million Automation Center (ATC) in Singapore to help its local clients deploy their automation and IoT solutions. OMRON expects the large manufacturing base in ASEAN to migrate to more automated processes[1], as well as Denka and Sumitomo Chemicals (Japan). All these vendors have built labs in Singapore to develop IIoT solutions and test the potential of their technologies in Singapore's factories.

Singapore is also home to many high-tech manufacturers in the electronics, aerospace, and petroleum industries. Many of these firms were early IIoT adopters. Now, many of these companies are also adopting digital manufacturing solutions based on robotics and AI.

Thailand is where most of the region's recent manufacturing IIoT spending exists. The Thailand 4.0 initiative guides the nation's IIoT spending. In line with the initiative, the Thai government announced plans to set up an Internet of Things (IoT) Institute in 2017. The goal: to develop IIoT in automation, robotics, aerospace, healthcare, and bioenergy industries.

Malaysia set up its National Internet of Things (IoT) Strategic Roadmap in 2015 to create a framework for long-term IIoT adoption. However, IIoT adoption in Malaysia is slow and unlikely to accelerate until manufacturers start showing healthier demand for digital manufacturing technologies. Malaysia's lower pay rates make many manufacturers reluctant to pay for technology that they feel they don't need yet. However, given the well-developed industrial

---

[1] McKinsey: Industry-4-0-Reinvigorating-ASEAN-manufacturing-for-the-future

infrastructure and ready manufacturing base, Malaysia has good potential to leapfrog the progress with identified use cases under vertical segments.

Indonesia's IIoT adoption is still in the slow lane, despite approximately 64 million IIoT sensors installed in the nation's factories. Constraints include perceived high startup costs, cybersecurity concerns, incompatibility with legacy systems, as well as the need for IIoT-trained operators and technicians.

In the Philippines and Vietnam, IIoT applications are also quite limited, because access to the internet and use of IIoT devices are still in their early stages. Installation of networks that support IIoT capabilities is accelerating in the Philippines.

A partnership of The Things Network and PacketWorx expanded strengthened ICT across the country by deploying more than 2500 LoRaWAN™ gateways. Manila-based Globe Telecom chose NB-IoT technology to support IIoT function in The Philippines. Narrow-band IoT technology was chosen for it's highly developed standard and its rapidly growing device ecosystem.

9.6   Big Data Analytics

First what is big data, and then what is big data analytics? Several years ago, you couldn't avoid all kinds of chit-chat about big data. Now the hype has cooled down somewhat, and it's much easier to gauge the value that it provides.

Big data used to be the term that data scientists gave to volumes of data too large for their traditional analytical tools and methods to handle. Fast-forward six years, from business intelligence to in-memory processing, data visualisation and more recently machine learning algorithms. These are just some of the tools and methods developed to gather, understand, and use structured and unstructured data of many kinds.

147

There are many ways to analyse the machine data that factory machines and devices generate by the truckload. However, the ways that engineers and business leaders put the data to work are far more interesting.

9.6.1 Aftermarket Services

In traditional ASEAN factories, few if any machines are connected to or stream data of any kind. In smart factories, however, manufacturers can combine big data analytics applications with machine learning algorithms and apply them to massive amounts of operations and customer data.

Analytics platforms gather, choose, and analyse structured and unstructured customer data coming from service contracts, ERP software, warranties, CRM information, and many other sources.

After collecting and cleaning data with proprietary algorithms, BDA platforms identify demographic or behavioural patterns to predict sales and service opportunities that are known to have high closure rates. Marketers can use what-if scenarios to test field service optimisations or aftermarket engagement campaigns without distracting business users from their work.

9.7 Improving Equipment Performance

To control production costs, manufacturers must get maximum value from their assets. Performance improvements, even small ones, can lead to higher revenue, fewer asset breakdowns, and lower capital costs. That's why process managers obsess on maintenance and continuous improvement of asset performance.

The data that IIoT devices send to company databases has great potential value. However, the speed and volume of incoming information can be overwhelming. The job of data analytics software is to capture, clean, and choose the most useful data, highlighting what's most important to a business. The whole idea is to reveal insights that can help technical and business leaders improve plant performance and achieve their business goals.

9.7.1 Making All Parts Traceable

Every manufactured part will have its own traceability code and digital quality certificate. Smart factory systems capture operational data such as production numbers and quantities, parts usage, machine status, and condition data. Traceability codes and identifiers follow each part from its fabrication to assembly, product launch, and perhaps to product recall or recycling.

9.7.2 Continually Monitoring Data

Just as data analytics software enables parts traceability, the high volume; high-speed capabilities of analytics make continuous monitoring of all production output possible, too.

AI agents and human workers have access to real-time data as products are made. Results of real-time data analytics appear to humans on dashboards or reports on the production line. In collaborative production environments, if any aspect of the product or process falls out of tolerances, either party can make changes to production variables. This monitoring process is continuous and can handle data on a monumental scale.

9.7.3 Defining Resource & Readiness Requirements

Using big data effectively requires data protection from internal and external security threats. Updating IT infrastructures, adding business initiatives, dealing with new regulations, and even dealing with employee digital devices, all make companies more vulnerable to threats on their information and data. The hyper-connected digital manufacturing environment requires companies to avoid or limit the risk of cybercrime and data loss related to natural disasters. Also, big data analytics requires lots of room for data storage and fast internet connectivity.

9.7.4   Big Data Analytics in ASEAN Manufacturing

Many ASEAN manufacturers have generated huge volumes of data for a long time. Although it's becoming easier than ever for enterprises of all sizes to use information to generate value, there's little data analytics activity in the ASEAN-6 manufacturing sector.

A February 2019 study by Analytics India Magazine describes how ASEAN companies have made analytics a key enabler of their ASEAN-6 operations. Manufacturing does not appear among the top seven sectors.

9.8   5G & Other Connectivity Technologies

Industry 4.0 discussions talk about connectivity in two ways. First, there's the general concept of connectivity as a function of an IT network and its connections to the internet. Second, there's connectivity technology, the software, hardware, standards, and best practices that enable data exchange and signals to travel throughout digital manufacturing systems.

Connectivity technology is an umbrella term for a wide variety of communications standards. These include wired (Ethernet, IO-Link, Modbus), short-range indoor wireless (Wi-Fi, Bluetooth, ZigBee), and outdoor long-range wireless (3/4/5G mobile, SIGFOX, Lora, and NB-IoT) communications.

Specific types of connectivity technology have roles in manufacturing-related use cases, including:

- Monitoring and tracking goods for real-time product status, condition, and location information — These cases use Wi-Fi,

Bluetooth, or Ethernet connectivity in the factory and warehouse and use 3/4/5G mobile on the road.

- Communicating with customers — Customers expect to be able to communicate with their vendors, anytime from practically anywhere. Unified communications technology enables customers to reach out by using voice, email, and instant messaging.
- Collaborating with product developers — Quick development and delivery of new and innovative products are essential for manufacturers who want to stay competitive. Connectivity technologies enable R&D collaboration between team members located anywhere in the world.
- Optimising asset management — Preventing downtime requires manufacturers to take a different approach to maintenance, one that uses data analytics to predict failure dates for production machinery. Keeping track of equipment life cycles and enabling proactive maintenance requires high-speed, high-volume data analytics. Transforming asset management data into actionable information requires a high-speed network.

9.8.1 Direct Machine-Business Communications

Discussions of Industry 4.0 manufacturing have prompted the new idea of machine-business connectivity. In the machine-to-business scenario, production data relevant to an enterprise's business is sent directly from the factory floor to the office of the company's business leaders.

Information content of this communications method depends on pre-set rules, which define the conditions and data content of the communications. This type of direct communications holds the promise of speeding up critical communications by cutting out the human middleman on the shop floor.

The IMF suggests the most important step in growing a digital economy is making internet connectivity universal and affordable. For businesses, fixed broadband internet service is essential because mobile internet services are often too slow and expensive.

Ultrafast broadband connectivity in the form of 5G technology provides essential support of AI, the IIoT, and other emerging technologies. Fixed-line broadband supports faster mobile internet services, enables businesses to cope with video streaming, and helps to manage supply chains with cloud-based computing services.

There's no strong correlation between broadband costs and Internet penetration in ASEAN countries. Broadband networks extend throughout high-cost nations such as Malaysia and Singapore. In less developed ASEAN countries such as Indonesia, as much as seven out of 10 people have no internet access at all.

However, broadband costs in higher-rate countries are likely to dampen adoption rates of emerging technologies, especially among ASEAN's small-to-medium enterprises.

9.9 Cybersecurity

As factory operations become more digitalised, their operations connect to many types of objects and assets. Despite the many capabilities that Smart Factories enable, they're open to a variety of potential cyber threats. The concern for safe computing in the factory makes some manufacturers reluctant to upgrade to emerging technologies. After all, who wants to give the cyber crooks "Open, sesame!" to their operations?

The pervasive connectivity of digital manufacturing systems makes cybersecurity more important than ever. The effects of cyberattacks can range from compromising the physical security of staff members to production downtime, damaged equipment and damaged brand reputations.

9.9.1   Types of Hacking & Cybercrime

There are so many ways to break-and-enter a corporate network that these functional descriptions are just a summary of an entire taxonomy of cyberattacks:

- **Stealing data** — With client details stored on CRM systems or customer accounts, hackers might try to find this information and hold it for ransom. Then again, they could do a straightforward data breach (with or without a diversionary, application-layer attack), steal the information, and disappear back into the internet.
- **Disrupting data access or operations systems** — Hackers can use scripts and hordes of bots (often IIoT devices) to overwhelm or shut down websites or take control of manufacturing processes. Bad actors can also interfere with the production process or tamper with products themselves.
- **Gaining intelligence for competitive advantage** — Every day, hackers and industrial spies steal intellectual property or other sensitive business information. The beneficiary can be the attacker (as in hacktivism) or a manufacturer's competitor. Because the IIoT is based on ubiquitous devices and connections, cyber-crooks with good programming skills can get access to information in many more ways than anybody wants to think about.

9.9.2 Production-related Threats

A quick glance at this list makes one fact clear —the number and variety of manufacturing cyberthreats are daunting. Here's a list of cyberattack targets that affect production equipment and processes (each category includes one or more types of attack):

- Connection between controllers (such as DCS and PLC) and actuators.
- Sensors (modification of measured values and states, their reconfiguration, etc.)
- Actuators (suppressing their state or changing their configuration)
- Information transmitted along networks
- IIoT gateways
- Operating panels, smartphones, remote controllers, and other devices & Safety instrumented systems

9.9.3 Cybersecurity Software Requirements

Artificial intelligence, the Industrial Internet of Things, big data analytics, and automated systems are not possible without three tiers of core infrastructure: secure fixed broadband connectivity, secure data centres, and secure cloud computing.

Now more than ever, enterprises will need to constantly monitor their IT operations and mitigate risks as needed. Integrated security begins with a customised cybersecurity audit and plan (completed in-house or with consultancy help). To protect all connected devices and environments on all fronts, secure smart factory environments should secure itself with these capabilities:

- Next-generation intrusion detection and prevention
- Application whitelisting
- Integrity monitoring
- Virtual patching

- Advanced sandboxing analysis

Machine learning-based behaviour analysis demands a risk-reducing security system and IT staff that stay informed about the latest cybersecurity threats and mitigation methods.

9.9.4   Cybersecurity & Industry 4.0 Readiness

Of course, these functional and resource items add a heavy load of time and money to other digital manufacturing readiness requirements.

The machine learning community has responded to these security challenges by providing high-speed, large-scale data and network monitoring capabilities. For instance, network-based solutions can help secure IIoT devices by preventing intruders from hacking into networks. Solving this authentication-based problem, however, requires finding, registering, and setting up a secure password for every device that accesses a network.

Machine learning engines that monitor incoming and outgoing IIoT device traffic can create a profile that determines the normal behaviour of the IIoT ecosystem. From there, detecting threats involves discovering data traffic and exchanges that do not fall within the expected behaviour profile.

9.9.5   Growing Exposure to Hacking & Cybercrime

Computer hacking and cyberattacks are on the rise throughout ASEAN and the world. Between March 2016 and May 2019, every ASEAN-6 country except Indonesia suffered a major data breach or cyberattack. In the ASEAN manufacturing sector, Toyota subsidiaries in Thailand and Vietnam experienced a series of data breaches in March 2019. Manufacturing operations were unharmed (customer information was exposed), but the brand's reputation was affected.

There's growing pressure for manufacturers to adopt integrated cybersecurity solutions and adopt software that monitors, protects, and even mitigates damage to vital equipment.

10 ASEAN: National Industry 4.0 Initiatives

"A good plan implemented today is better than a perfect plan implemented tomorrow."

— *George Patton*

ASEAN has no regional Industry 4.0 initiative. Instead, several ASEAN member countries have launched their own initiatives. This section describes the Industry 4.0 initiatives of six ASEAN nations with vigorous manufacturing sectors.

10.1 Singapore

Singapore raced through Industrial Revolutions 1.0 through 3.0 in less than 50 years. Pro-business policies, competitive taxes, a robust intellectual property regime, and an educated, reliable, and adaptable workforce made this remarkable progress possible. Although manufacturing generates a fifth of Singapore's GDP, the nation isn't recognised for its manufacturing base.

Nevertheless, the country is using Industry 4.0 technologies to re-establish its position as a preferred location for manufacturing innovation and the development of new ideas and solutions. The country's Industry 4.0 adoption effort is guided by several initiatives and policies. These initiatives include:

10.1.1 Smart Nation

Throughout decades of rapid economic growth and urbanisation, Singapore's policymakers have addressed increasingly complex policy problems by using emerging technology solutions such as data analytics. The Singapore government gathered a series of e-government programs under the banner of the Smart Nation Initiative and launched it in late 2014.

This nationwide and government-wide effort helps Singaporeans digitise the nation's policy processes and urban environment. The initiative groups specific concerns, such as urban living, transport, health, and start-ups and businesses. Industry 4.0-related initiatives include the Smart Nation Innovation and Data Innovation Program Office programs.

Virtual Singapore Program (Digital Twin of Singapore): Virtual Singapore is a dynamic three-dimensional (3D) city model and collaborative data platform, including the 3D maps of Singapore[2]. This project is championed by the National Research Foundation (NRF) at a cost of $73 million. The major capabilities of virtual Singapore are:

- Virtual Experimentation
- Virtual Test-Bedding
- Planning and Decision-Making
- Research and Development

For instance, the planner can use the app to assess the solar potential of certain buildings through simulated placement of solar panels, and develop solutions to improve accessibility for families, the ageing population and people with disability!

Industry 4.0 Accelerator Program — Launched in April 2017, a collaboration of the Economic Development Board and the Boston Consulting Group. With this program, BCG forms a partnership with the Singapore Economic Development Board. BCG offers its capabilities and experience to help ASEAN manufacturers improve

---

[2] https://www.nrf.gov.sg/programmes/virtual-singapore

their manufacturing operations by introducing advanced technologies, new digital processes, and the latest management practices.

Digital Economy Framework for Action — Launched in May 2018, the DEFA documents the digital economy pillar of Singapore's Smart Nation initiative. The program goal is to help Singaporeans become the world's leading digital economy. Key aspects of the plan focus on collaboration between government agencies, business partners (users), and IT platform providers.

The national government also launched a first-of-its-kind tool, the Singapore Smart Industry Readiness Index. This decision-making aid provides a systematic and comprehensive way for manufacturers to determine the potential impact of Industry 4.0 on their operations.

10.1.2 Drivers of Industry 4.0 Initiatives

Singapore's manufacturers are grappling with rising operations costs, stagnant productivity, and a domestic labour crunch created by an ageing workforce. Government representatives acknowledge the need to improve its manufacturing model to one that offers more innovative, high-value production.

Authors of Industry 4.0 initiatives view digitalised technologies as enablers, which will help Singaporean manufacturers:

- **Find new revenue streams** — Government initiatives will help companies transform business models and expand into new areas of business.
- **Tackle workforce shortages** — Industry 4.0 technologies are expected to improve productivity and optimise manufacturing processes.
- **Serve more global customers** — Emerging technologies will enable manufacturers to provide higher-value products from manufacturing facilities in Singapore.

- **Create value-added jobs** — Boosting Singaporean society as a whole.

10.1.3 Potential Barriers to Ind 4.0 Adoption

In its national Industry 4.0 strategy, Singapore's government has earmarked significant financial support and other resources for investment in Industry 4.0 programs. These include financing for manufacturing R&D, developing Industry Transformation Maps (ITM), and strengthening workforce skill sets. However, obstacles to Industry 4.0 adoption remain.

- **Need for more designer and managers** — The nation's educators turn to teach, train, and encourage future manufacturing and engineering professionals to develop modern skills and a flexible, think-for-yourself attitude.
- **Developing technical innovators and entrepreneurs** — Much of the value of Industry 4.0 technology will belong to innovators and entrepreneurs. These are the folks who convert innovations into manufactured goods. Few Singaporeans have the training and mind-set needed to generate manufacturing value and retain it in the country. Conventional technical and business thinking will make it hard for traditional manufacturers to compete.

If a developed country like Singapore has potholes in its road to Industry 4.0, what's the status of another nation experienced in global manufacturing markets? Malaysia provides a good example of the progress and challenges of a mid-income nation wanting to be more.

10.2 Malaysia

The Malaysian economy is the third-largest in Southeast Asia and the 38th largest economy in the world. The country's competitiveness is strengthened by high labour productivity, knowledge-based industries, and adoption of cutting-edge manufacturing technologies. According to the Global Competitiveness Report 2017, the Malaysian economy is the 23rd most competitive country in the world.

Malaysia's key challenge is how to stay competitive in a world of lightning-quick technological and business change. Like many other ASEAN member states, the Malaysian government views Industry 4.0 adoption as an effective way to build up its technological and industrial muscle. Its Industry 4.0 initiatives can be found in several strategies and plans, which include:

- National Policy on Industry 4.0 (MOSTI)
- Industry 4.0 Policy Framework for the Manufacturing Sector (MITI)
- Study on Future of Manufacturing: Industry 3+2 (MIDA)
- National Internet of Things (IoT) Strategic Roadmap,
- Digital Malaysia: Malaysia Digital Economy 2017
- Industry4WRD

10.2.1 Industry4WRD Drivers & Objectives

Malaysia's National Policy on Industry 4.0, the Industry4WRD plan, builds on ideas taken from earlier documents. As the country's definitive Industry 4.0 initiative, Industry4WRD describes the nation's plan to transform Malaysia's manufacturing and service sector in the next 10 years

Malaysian government officials view Industry 4.0 as an opportunity to become a strategic partner with smart factory operators and providers of related services in the Asia-Pacific region. Closer to home, Industry 4.0 adoption is recognised as an enabler, which can help manufacturers —especially SMEs— to gain or keep a competitive advantage.

The Industry4WRD plan recognises 10 technology enablers, namely autonomous robots, augmented reality, big data analytics, cloud-based computing, system integration, additive manufacturing, Internet of Things devices, cybersecurity, simulations, and artificial intelligence. In the plan's framework, manufacturers will use these enablers to meet these general objectives:

- Enhance global competitiveness.
- Strengthen the growth and productivity of the Malaysian manufacturing industry.
- Create new-generation jobs and reduce dependence on foreign labour.
- Sustain high levels of foreign and domestic investment. The Industry4WRD plan also sets out these specific performance targets:
- Increase per capita manufacturing productivity by 30 per cent.
- Increase the manufacturing industry's contribution to the national economy to RM392 billion (up from RM254 billion, a 54 per cent increase).
- Improve Malaysia's Global Innovation Index ranking from 35th to a position in the top 30.
- Increase the manufacturing industry's segment of highly skilled workers from its current value of 18 per cent to 35 per cent.
- The Industry4WRD plan describes a collaboration of stakeholders from government, manufacturing, and education sectors. In the plan, stakeholders would work together to carry out these high-level objectives:
- Strengthen digital connectivity.
- Enhance the capabilities of the existing workforce.
- Help manufacturing workers develop new work-related talents and skills.
- Improve data integrity methods and standards.

162

- Strengthen research, innovation, and technological development programs and activities.

Initiative participants will carry out these objectives through targeted action plans related to funding, infrastructure, regulations, skills, talent, and technologies.

10.2.2   Readiness, a Top Priority

Adopting Industry 4.0 technology and ideas can't happen until each manufacturer is ready. That's why clarifying each manufacturer's readiness is a fundamental part of the Industry4WRD plan.

To provide clarity and support to manufacturers on their journey to Industry 4.0 adoption, the Malaysian government published its Industry4WRD Readiness Assessment Guidelines. This self-diagnostic tool helps Malaysian manufacturers.

10.2.3 Obstacles to Improved Manufacturing

The Malaysian government provides incentives to manufacturers who adopt automated equipment, ICT hardware and software, and smart manufacturing equipment. These inducements include tax Incentives, soft loans, and automation capital allowances.

Malaysian manufacturers are relatively well-positioned to improve their manufacturing technology and business processes. However, there are still substantial gaps in the country's manufacturing and supply chain capabilities. These include:

- Low rates of digital adoption (~20%) especially among SMEs and limited use of automation by manufacturers — Most Malaysian manufacturers automate fewer than 50% of their operations.

163

- Urgency for an integrated, digital approach to data gathering — Often, Malaysian manufacturers still use spreadsheets and analyse data manually.
- Concerns about cyber threats, due to increased connectivity of Industry 4.0 technologies (especially IIoT devices) — This makes manufacturers reluctant to expose their data to security and privacy vulnerabilities.
- High demand for workers and professionals with required skills and knowledge — This shortage is most pronounced for specialists in IIoT, robotics, and AI.
- Concerning costs and payback periods for Industry 4.0 projects — Industry 4.0 technologies and processes generally cost more and take longer to pack back than those in legacy manufacturing environments. These trends cause concerns among manufacturers, who are often don't see enough short-term value to adopt new tech and methods.

Malaysia is another ASEAN-6 nation that benefits from a history as a successful manufacturer of low-to-mid-value products. The challenge for Malaysians is to summon their organisational abilities and political will to use government incentive programs and take the plunge to adopt Industry 4.0 thinking and technology.

10.3 Thailand

Thailand 4.0 prioritises developing a value-based economy that can help the emerging economy avoid the middle-income trap. To accomplish this goal, manufacturers must compete more effectively with rival countries within and outside ASEAN. Their strategic framework emphasises using innovation and technological advancement to preserve Thailand's competitive edge.

10.3.1 Initiative Drivers & Objectives

At present, Thailand is facing the challenges associated with the capabilities to progress in many of the plan's target industries. The success of the Thailand 4.0 strategy will depend on the country's attracting foreign direct investment.

In February 2017, legislation expanded Thailand's Industry 4.0-related incentives for companies that use digital technology as well as advanced production methods and materials. These incentives include:

- Income tax exemptions for periods up to 15 years.
- More options for foreign shareholders.
- More land ownership options for investors.
- Exemptions for import duties.
- R&D subsidies valued at Bt10bn (US$300 million). These and other incentives included in Thailand 4.0 rivals all past industrial policy concession packages.

10.3.2 Obstacles to Successful Ind. 4.0   Adoption

Analysts recognise that Thai Industry 4.0 efforts will significantly improve the manufacturing efficiency and worker productivity that government and business stakeholders crave. However, shortcomings in the plan and persistent problems in the country's finance and education sectors make the transformation of Thai society uncertain.

Heavy on intentions, light on tactical details. The plan makes the government's intentions clear. Focus on 10 enabling technologies to make Thai manufacturing and supply chain processes more efficient and workers more productive. However, the plan lacks important details—deadlines, performance indicators, and clearly defined accountability measures—that make initiatives successful. Specifically, the plan lacks:

- A step-by-step process that will lead users to their objectives.
- Critical success factors that support plan objectives or key performance indicators that indicate how well businesses are doing in the transformation process.
- A single-focus stakeholder or agency that would oversee planning. Directing industries by defining what they will deliver during the plan's three-to-five-year term.

10.3.3 Societal Barriers to Success

Thailand's digital manufacturing developments reflect many challenges of its ASEAN-6 neighbours.

Investment in R&D — According to recent World Bank figures, Thailand spent merely 0.6 per cent of its GDP in 2017, compared with Malaysia's 1.3 per cent and China's 2.1 per cent.

Ongoing shortage of skilled labour and professionals — In Thailand, skilled labour is scarce. It's also a major barrier to FDI and an obstacle to increased R&D spending.

About half of all secondary-school graduates further their education - In a recent report, only half of all students surveyed showed minimum proficiency in reading; only 46% of students met minimum standards in mathematics. This, a slowly growing labour force, has compounded the shortage of skilled labour.

Government contribution to national R&D effort — Designed to stimulate and direct R&D activity throughout the country, the Thai National Science and Technology Development Agency (NSTDA) has yet to make a significant contribution to R&D. Thai analysts have called for a major overhaul of the country's key research body and federal laboratory.

A shortage of a skilled, knowledgeable workforce will likely slow down Thailand's Industry 4.0 adoption process. However, Thai manufacturers need to also contend with vague Industry 4.0 program objectives and uneven quality in the education of future manufacturing workers. Effort for the nation's manufacturers to figure out national goals to get the workforce needed to make Industry 4.0 work for them will be critical in the coming years.

10.4 Indonesia

In April 2018, Indonesian government leaders released Making Indonesia 4.0. This roadmap documents the strategies that Indonesia's private and public sectors must implement to benefit from the use of Industry 4.0 technologies.

10.4.1 Initiative Drivers & Objectives

Manufacturers and government leaders plan to use progress in Industry 4.0 development to achieve its core aspirations: to revitalise the country's manufacturing sector and regain its standing as a global manufacturing powerhouse. Government representatives expect Making Indonesia 4.0 strategies to help the nation hit these goals:

- Become one of the world's 10 biggest economies by 2030.
- Return to the manufacturing industry's former net export rate of 10 per cent.
- Double the labour productivity rate over labour costs
  Allocate 2 per cent of the national GDP to R&D and technology innovation, which would be seven times the current investment rate.

To reach these goals, initiative stakeholders focused on a list of high-priority capabilities:

- Improve the quality of training and education for workers and professionals in the manufacturing sector.
- Empower micro-, small-, and mid-sized companies as well as enterprises.
- Provide incentives for technology investment
- Establish a national innovation network.
- Attract foreign direct investment.
- Harmonise regulations with current technology and innovation policies.
- Build a national digital infrastructure.

Barriers to Industry 4.0 implementation Indonesia aspires to a future that restores its past role as a manufacturing hub and value-generating powerhouse. However, success requires a monumental effort to fill many gaps in the country's manufacturing and other infrastructures.

Human resources — Indonesia must develop its technical and professional workforce. The nation will have plenty of workers for the next 15 years. However, Ministry data indicates that unskilled labour dominates the Indonesian workforce (60.2 per cent of a total 125.4 million workers). Only 11.6 million workers hold a bachelor's degree certificate.

Use of Industry 4.0 technologies is predicted to eliminate low-skilled jobs and generate new job opportunities that require technical competencies. So, it's necessary to develop technical knowledge and skills of Indonesian manufacturing workers and professionals. Not enough young students in Indonesia have the right skills, particularly in the technical, scientific, and engineering fields. Adoption of Industry 4.0 tech would require aggressive reforms to Indonesia's public and higher education systems.

Insufficient capital — Industry 4.0 projects are high-tech, capital-thirsty beasts. The Indonesian banking industry has not focused on financing these projects because Industry 4.0 technologies are relatively new and are high-risk propositions.

Educational and research infrastructure — The nation's education budget is limited to US$114 per student. Indonesia's government has a very tight R&D budget, about 0.3% of GDP. Government officials admit that the nation's R&D budget must rise to at least 2 per cent for Indonesia to become competitive in advanced digital manufacturing.

Digital Infrastructure — Indonesia is limited by its digital infrastructure. The 4G cellular phone infrastructure that people use is not ready for 5G technology. The average fibre optic speed is low (less than 10 Mbps), and cloud-based infrastructure is also limited.

Industrial support infrastructure — Hitting initiative program targets requires the availability of abundant, cheap, and reliable electricity. Reliable internet services with substantial bandwidth and extensive coverage are also on the list of must-have resources. Manufacturers will also need a modern logistics infrastructure and data centres with safe, secure, high-volume data storage. Only few of these is on the development horizon.

10.5 Vietnam

In May 2017, Vietnamese Prime Minister, Nguyễn Xuân Phúc, announced the government's new Industry 4.0 strategy. The Prime Minister's rationale for this new policy was straightforward. He wanted Vietnam to improve its existing infrastructures and invest in innovative, new technologies that would promote economic growth. The goal: to seize new economic opportunities and minimise possible social disruption on the horizon.

Vietnam 2025, the umbrella term for the government's national Industry 4.0 strategy, focuses on:

169

- The government creating specific laws, regulations, and policies that enable businesses to learn and develop new technologies and business models quickly.
- Companies upgrading outdated work practices that aren't consistent with Industry 4.0 processes and methods.
- Collaboration between government, educational, and business stakeholders.
- Investment in emerging technologies.
- Skills training and degree development programs.
- Continued development of an innovation ecosystem — Vietnam 2035

Vietnam 2035 is Vietnam's Industry 4.0 vision document. Its content describes how emerging technologies will be applied to all industries and fields. These policy efforts have a special urgency, created by worrying economic forecasts and demographic change.

10.5.1 Initiative, Opportunities, & Disruption

Over and over, the same phrase appears in Vietnamese government speeches, blog posts and newspaper articles. Like their ASEAN-6 peers, Vietnamese government officials are trying to foster business opportunity as they ride the tiger of rapid social, technological, and business change. Vietnamese government officials and business leaders must deal with three pressing drivers of change.

A longer global reaches for Vietnamese manufacturers — First, a positive driver. As an emerging economy, Vietnam has a lot to gain by using Industry 4.0 technologies and processes.

We're all familiar with the efficiency and cost-control improvements that support the rationale for Industry 4.0 adoption. However, the hyper-connectivity of manufacturers with suppliers, vendors, customers, and partners provides seldom-mentioned benefits to members of Industry 4.0 ecosystems. Simply stated, the more closely Vietnamese manufacturers adopt Industry 4.0 technologies and ideas, the more potential customers they can reach.

Erosion of Vietnam's competitive advantage — The move to Industry 4.0-centred development policy marked a significant change from Vietnam's previous economic growth model. This relied heavily on competitive advantage in markets for natural resources and lower-value, labour-intensive manufacturing. As these markets shrink, business leaders and government officials must look for something to fill the gap.

Changing workforce demographics — Currently, Vietnam has a workforce that's young and adaptable to changing technology and work practices. However, the Vietnamese government is taking the long view, and what they see is worrisome. The nation's workforce is ageing and slowly decreasing. It's expected that by 2050, 20 per cent of the Vietnamese population will be more than 65 years old. So, the current shortage of working-age Vietnamese with modern manufacturing knowledge and skills will get worse over time.

10.5.2 Barriers to Industry 4.0 Adoption

Vietnam was relatively slow to form a national Industry 4.0 strategy. Late arrival to Industry 4.0 awareness and the lack of specific targets and clear program definitions suggest a problematic future for the initiative's success. Here are five challenges that won't easily go away.

Relatively low Industry 4.0 awareness — There's concern among government officials that business community awareness of Industry 4.0 is still low. Currently, it remains uncertain to tell whether the low level is a matter of awareness or interest.

Generally low Industry 4.0 readiness — Even if more Vietnamese manufacturers were aware and supportive of Industry 4.0 ideas, many are not ready to adopt it. Often, they lack the communications and technical infrastructures and experience using IT apps in the industrial sector. Simply choosing hardware and software will not work magic. First, manufacturers must prepare their factory and office environments for Industry 4.0 processes.

The need for supportive government policies — If Industry 4.0 is going to succeed in Vietnam, the government must first ensure a business environment that enables innovation and modern business practices.

Currently, Vietnam is suffering from a start-up brain drain. Resourceful, risk-tolerant start-up owners from Vietnam are moving to the more business-friendly environment like Singapore. This is just another example of the need for more supportive government policies and practices. Government rhetoric claims to be aware of this problem and intending to make improvement.

Too few university graduates in STEM and management careers — The Vietnamese people are in dire need of more, higher-quality university degree programs. In 2017, only 8 per cent of the labour force had a university education. In future Industry 4.0 manufacturing jobs, there will be more open positions than Vietnamese technical professionals and engineers to fill them. Filling these jobs with foreign candidates will add new problems to manufacturers' challenges.

Not enough investment in the R&D stage of commercialisation — Like many of its peers, Vietnam calls for more human resource infrastructures. These networks of knowledge, technology, and business experience would enable and accelerate the commercialisation of good ideas into manufactured products and services within the country.

10.6 Philippines

The Philippines has experienced vibrant economic growth in recent years. However, the country is recognised as a legacy economy, with a strong production base but a low level of readiness for Industry 4.0 manufacturing.

Filipino manufacturers and government officials acknowledge that entrepreneurship can stimulate the sector's competitiveness and productivity. However, the Philippines' position in global innovation is relatively low (79 of 134 globally).

The country's Inclusive Innovation Industrial Strategy (i3S) is designed to increase innovative and competitive manufacturing. Developed by the national government's Department of Trade and Industry's Board of Investments, the plan focuses on developing innovative and globally competitive manufacturing businesses. The report focuses on value created by domestic and global value chains.

10.6.1 Competitiveness, Innovation, & Productivity

The strategy's central idea is the relationship between a healthy, competitive economy, innovation, and manufacturing productivity. The i3S plan guides manufacturers to establish a market environment, which leads to more competition. This, in turn, spurs innovation and productivity growth. Pillars of the i3S plan include:

- Developing innovation and entrepreneurship.
- Increasing manufacturing capacity.
- Developing human resources.
- Developing micro-, small-, and medium-sized manufacturing businesses
- Making it easier to do business with and invest in manufacturing businesses.

10.6.2 Drivers & Objectives

The I3s plan focuses on innovation and entrepreneurship in electronics, automotive, aerospace, chemicals, and agribusiness.

Government officials and business owners recognise the need to move from legacy manufacturing to higher-value production.

However, changes to more advanced manufacturing require automation, which will put many Filipino manufacturing jobs at risk. In the i3S plan, the need to compete more effectively and develop high-value manufacturing capabilities leads stakeholders to engage in these activities:

- Promoting collaboration and closer coordination with government agencies.
- Creating a regional network of regional Inclusive Innovation Centres. Participants will include universities, R&D labs, science and technology parks, incubators, and investors.
- Improving worker productivity by developing skills in science, technology, engineering, and mathematics.

10.6.3 Roles & Responsibilities

In the i3s framework, the private sector is the primary driver of implementing Industry 4.0 adoption. The government coordinates policies and provides the necessary support to remove obstacles to development and grow the manufacturing industry. Government stakeholders contribute to the plan by:

- Securing funds for the Filipino Industry 4.0 Innovation Program.
- Relaxing regulations related to procurement, employment of foreigners, and innovation-related services.
- Working with industry representatives and educators to create policies and training programs that match rapidly changing technology and business requirements.

- Working with university educators to guide the formation of courses and programs that will advance innovation and entrepreneurship.

In the Philippines, the government's relatively weak role in implementing Industry 4.0 adoption is unusual among ASEAN-6 members. The difficulties the nation faces, however, are all too familiar.

10.6.4 Barriers to Industry 4.0 Adoption

The Philippines has decades of experience as a legacy manufacturer. Now, the nation wants to upgrade the value of its manufactured goods. Taking this huge step requires resources and capabilities that haven't been available previously.

The Philippines lacks a reliable internet infrastructure, as do many of its ASEAN neighbours. However, one of the obstacles to successful adoption of Industry 4.0 technologies lies in the Filipino education system.

Problems in the education sector — Given the critical role of education in Industry 4.0 development, Filipinos will have a challenging time building the manufacturing skills and production capabilities they need. Low levels of research funding, a lack of skills training, and little progress in developing technical degree programs are the culprits.

In the Philippines, government support of academic research is low, only half of one per cent of the nation's 2018 GDP. China, the United States, and other manufacturing competitors in the European Union spend about 2 per cent annually. When it comes to funding, education and training institutions get short shrift generally — annual spending for Filipino primary and secondary students averages about USD 114 per student.

Lack of Skills among manufacturing workers — There's also an outstanding gap between Filipino schools and industries. There is little connection between businesses, universities, and training institutions. These silos of education make it difficult to identify and develop the higher-value skills that might make Filipino workers less vulnerable to job loss by automation.

Universities and businesses are also in sore need of ongoing collaboration. Educators have yet to develop courses and programs in STEM careers and engage in applied manufacturing research to build technical degree programs.

The i3s initiative is a good start and the many to-do lists are still work-in-progress thus political will to make things happen is critical.

10.7 Comparisons of ASEAN Initiatives

When you read through the descriptions of Industry 4.0 initiative drivers, intentions, and obstacles, it's easy to see a lot of overlap. ASEAN-6 countries generally share the same problems of modernising their education systems, training workers with useful skills, and improving infrastructures so vital to Industry 4.0 readiness.

The main difference, however, is the role that different national governments play in the Industry 4.0 adoption process. For example, Singapore's government is comfortable and energetic in its leadership role in modernising its manufacturing base. Government agencies offer monetary incentives and software tools. They also support many opportunities for manufacturers and other stakeholders to build relationships that are essential to digital transformation.

Other nations, such as Vietnam, are slower to join the transformation process. SRVN initiatives take the form of policy intentions delivered in speeches and newspaper articles. To date, Vietnam needs to establish authoritative, single-source strategy document, which contains targets, deadlines, and specific commitments.

11 Getting adoption off the ground

In print or on screen, everywhere you turn to listen or read the message is clear: Industry 4.0 technology enables substantial economic and social benefits. These articles, blog posts, and podcast make one thing very clear: ASEAN manufacturers have no time to waste. The process of adopting Industry 4.0 technology and ideas must begin now.

11.1 Adoption of Digital Manufacturing?

Do you wonder who and what drives the adoption of manufacturing production, process, and information technologies in ASEAN?

Technology change and macroeconomic forces within and beyond ASEAN member countries drive the case for Industry 4.0 technology adoption. Government agencies are eager to provide financial inducements (the carrot), while the spectre of strong intra- and international competition provides the stimulus to take a chance on change (the stick).

This chapter reviews general requirements and country-specific responses to starting and Industry 4.0 adoption throughout ASEAN-6 countries: Indonesia, Malaysia, Singapore, the Philippines, Thailand, and Vietnam.

11.1.1 ASEAN's Ecosystem for Ind. 4.0 Adoption

Industry 4.0 adoptions in ASEAN depend on the region's system of technology users, national and regional institutions, and the economic and social trends that drive technology use.

The roles and relationships of government agencies, businesses, and educational institutions determine how rocky or smooth the transition to smart manufacturing will be. In this chapter, our goal is to get clear about what it takes to integrate converging technologies with existing manufacturing technologies, processes, and ideas. First, it helps to identify the players and drivers of Industry 4.0 adoption. We'll look under the hood of ASEAN institutions to discover:

- Government, education, and business groups, which define and drive manufacturing technology adoption.
- How these groups participate and relate to each other in Industry 4.0 adoption efforts.
- The effect that these relationships have on converging technology use and manufacturing business success.
- Barriers to Industry 4.0 adoption and the steps that these groups can play to accelerate the adoption process.

What drives Industry 4.0 adoption in ASEAN? It's easy to discover why manufacturers want to adopt Industry 4.0. One need only open a newspaper or consultancy white paper to see it in print. ASEAN manufacturers and governments want to expand on—and in some cases, recapture— their status as a global manufacturing centre.

Achieving this goal, however, will not be easy. Regional and global manufacturing environments are different, and much of worked well in past decades won't work now.

11.1.2 A New Kind of Manufacturing Workforce

One of the strongest drivers of Industry 4.0 adoption is the need for a new type of manufacturing workforce. Some countries, such as Singapore face a shrinking and ageing workforce. Other nations such as Vietnam have plenty of workers, who required higher knowledge and skills to operate advanced machinery and design products with new methods and tools.

On the short term, ASEAN-6 manufacturers, especially SMEs still need line workers with today's manufacturing skills. Longer-term, however, manufacturers will need employees who can work with robots, design using new materials and methods, and operate technology. These requirements translate into a need for highly trained operators, engineers, and technical professionals with backgrounds in electronics, material science, and medicine.

11.1.3 Technology requirements of ASEAN Region

Need for advanced technology — Another major driver of Industry 4.0 adoption is the need for advanced technology, which can make manufacturing processes more efficient and—manufacturers hope—more competitive.

Some of these technologies, such as IIoT devices are familiar but lend valuable new capabilities and efficiencies to manufacturing and supply chain processes. Others such as additive production (3D printing) offer new design and production opportunities to manufacturers.

Manufacturers throughout the region have an ongoing need for advanced production, process control, and information technologies. As examples, here are two manufacturing examples that illustrate these requirements: the automotive and textile industries.

Automotive industry — Here are just a few of the many methods, tools, and machines that modern automotive manufacturing requires:

- Production methods used in electric and hybrid electric vehicle production.
- Abilities to fabricate and design parts with advanced, lightweight materials.
- Automated assembly. Production-line robots are becoming smaller, cheaper, and easier to install. Moreover, they're more adaptable and capable of working with people.

- Internet-connected sensors and devices. Vehicles that drive themselves and notify roadside help if they are damaged — enabled by wireless, internet-connected technology.

Textile industry — Consumer demand will start driving clothing and textile manufacturers to use advanced materials, new production methods, and digital technology.

For example, Advanced materials such as nanoparticles used to produce odour-free, waterproof, UV-blocking, or antistatic clothing. Advanced CAD software used to customise apparel. Finally, medical body scanners embedded clothing to produce "smart apparel," which gathers the wearer's biometric data.

What will provide the production and process control capabilities that manufacturers need to become and stay competitive? That's where leaders from the public and private sectors come into the picture.

11.2 ASEAN Government Initiatives & Schemes

In the past, the policies of ASEAN national governments have served as powerful influencers and directors of technology programs and initiatives. Currently, national governments of Industry 4.0 early adopters can play many important roles in getting the process started.

11.2.1 Technology Leadership

ASEAN governments are taking active roles in encouraging Industry 4.0 technology adoption. The national governments of Singapore, Malaysia, and Thailand provide examples of technology leadership. They provide the coordination, promotion, advocacy, and strategic planning support that technology adoption requires.

The essential ingredient in this complex stew of thoughts, deeds, and intentions is a clearly identified focal point. Nothing can begin until someone answers several important questions.

Who's in charge of Industry 4.0 adoption? This person (ideally, one individual) is the focal point of Industry 4.0 adoption in each ASEAN government. Ideally, leadership would involve some degree of involvement of the head of government or high-ranking staff. Regardless of who is in charge, the individual should be the final arbiter, who defines Industry 4.0 policy and answers policy-related questions.

Ideally, ASEAN would work within a regional framework. This approach would describe adoption strategy and implementation for all ASEAN countries. However, considering the varying levels of business development, technology adoption, and economic strategies, national policy frameworks are the pragmatic approach.

The most effective national documents would provide a clearly written, easy-to-access specification, which defines who does what.

11.2.2 Becoming Technology Thought Leaders

Successful digital transformation requires more than using new hardware and software. It requires new ways of thinking about technology use. Thoughtful reviews of Industry 4.0 adoption in ASEAN refer to the constraints of legacy technology and the dangers of legacy thinking.

Ideally, national governments would take the lead in encouraging manufacturers to become more risk-tolerant. Encouraging Industry 4.0 adoption is the key to modernising ASEAN manufacturing. In Industry 4.0 ecosystems, governments are responsible for providing the legal and administrative environment that encourages the adoption of new technologies. Laws and rules must be clear enough and specific enough to reduce the risks of long-term decision making and they must be flexible enough to avoid red tape.

11.2.3 Creating a Friendly Environment

Developing a productive, job-ready workforce that's able innovates and think critically is a prime objective of Industry 4.0 initiatives. What can governments do to create a policy environment that encourages implementation? They can offer a wide variety of policies, incentives, and support services.

- **Create technology roadmaps** — To start, national governments can help manufacturers get a clear picture of obstacles that delay Industry 4.0 adoption. Developing national Industry 4.0 roadmaps is another helpful way to speed up the acceptance of digital manufacturing. Each ASEAN member will have its own way to map the way forward for its manufacturers. Some nations have already started this process.

Governments such as Singapore, Indonesia, and Thailand are creating national roadmaps for adopting Industry 4.0 technologies and ideas. Also, Singapore has produced its national Smart Industry Readiness Index, a first-of-its-kind self-assessment tool. Each roadmap helps manufacturers decide where they are in the Industry 4.0 adoption process and how they might move forward.

- **Promote Skills development schemes** — Supporting nation-wide, skills development projects is another example of a helpful government programme. Singapore's SkillsFuture initiative, a lifelong skills development program, is a good example of Industry 4.0 education and training programs. Guided by Singapore's Future Economy Council, this initiative combines talent from government ministries, corporations, academic institutions, and labour groups.
- **Offer tax incentives to boost innovation** — Governments can support companies to make the leap to Industry 4.0 adoption by offering preferential tax policies.

- **Create special funds to encourage industrial transformation**
  — Programs such as Italy's Enterprise 4.0 national plans is an example of a government-funded, business improvement scheme.
- **Encourage regional cooperation** — ASEAN governments could encourage regional exchange and cooperation in the form of Industry 4.0 pilot and demonstration projects.
- **Support training and talent initiatives** — ASEAN governments could also encourage pilot and demonstration projects that develop top-level design and engineering talent. As the references to life-long learning and skills improvement programs show, government policies alone cannot assure success.

Schools and academic institutions also have important roles in improving and modernising manufacturing skills and knowledge.

11.3 ASEAN Educational Institutions

In most ASEAN nations, Industry 4.0 adoption is starting slowly and progressing at different speeds. However, the trends that drive transformation are already at the region's doorstep, demanding change. The work world that is here (or will be here soon) requires workers who can:

- Help fill the region's productivity gap.
- Work with computers, machines, and other people; communicate with each other; and perform complex tasks in manufacturing environments.
- Adjust to job loss and job fragmentation caused by automation and other advanced technologies. If you picture vocational or university education as it was 10 years ago,

throw that image away. It does not apply to the world of Industry 4.0. The topics, who teach them and how they are taught, are all being transformed.

11.3.1 Higher Education

Futurists are predicting that intelligent machines will take over all or part of routine monitoring and analysis tasks in professional jobs. These predictions include tasks in entry-level positions within medicine and the law, for example. While robot physicians' that take over direct medical duties from humans are not on the horizon, the use of robotics as supplemental tools in clinical settings is a wakeup call to educators everywhere.

Industry 4.0 will require deep changes in educational content, teaching methods, and program management, even for highly skilled professionals. However, there are many things ASEAN institutions of higher education can do to prepare manufacturing professionals for the future.

Most of these changes are big-picture items, which require planning, resources, and political will to accomplish. However, these goals should alert educators to the capabilities and resources needed to support tomorrow's manufacturing professionals:

- Develop educational content and programs that challenge, reward, and support highly creative graduates, who can think critically.
- Help students in all disciplines to acquire digital and data literacy during their studies.
- Consider the increased use of blended, practice-oriented, project-based, and scenario-based learning.
- Create curricula and post-graduate programs that make lifelong learning a permanent part of professional life.

- Consider new partnerships between educational institutions and ASEAN manufacturers.

11.3.2 Skills Development & Training

New technology-based work skills and training are becoming essential to non-professional employees, too.

Vocational educators and trainers are also actors in the ASEAN Industry 4.0 adoption system. They will operate in a volatile business and social environment, which includes:

- A demographic explosion of job seekers entering regional job markets — This trend is expected to occur when many existing jobs are eliminated or changed by automation and AI.
- Shrinking tax income for national governments caused by fewer manufacturing workers.
- Need for more investment in workforce retraining. As some legacy manufacturing jobs disappear, workers will have to engage in the process of continuous retraining and skills development.

What educational and training institutions can do? — Scarcity of skilled workers is a critical bottleneck in many ASEAN countries. Getting the right manufacturing skills and training to workers in manufacturing hubs is a prime challenge to educators. There are several ways that educators can make this happen.

- **Teach the capabilities that local manufacturers need** — Education and training institutions should emphasise vocational education and provide other tailored training programs. That means adapting their programs to provide a specific skill that makes students job-ready for factory work.

185

- **Follow successful examples from other countries** — Other countries have managed the technology and labour challenges that ASEAN educators face today. Successful models combine formal learning, vocational training, and practical work experience. The proven apprenticeship systems of Germany and Switzerland are good examples of this approach. In these countries, governments, educational systems, and employers work together to build the skills and experience that local labour markets need.

- **Collaborate with governments and private sector companies** — Regional and international collaboration has already started to develop in vocational education programs within ASEAN. Such as the Global Apprenticeship Network, which designs and implements apprenticeship schemes in ASEAN countries. These programs combine the efforts of educational institutions and multinational companies to modernise the region's vocational education. Current and ongoing pledges of GAN13 corporate members will have a positive effect on more than 9.3 million young people through 2020.

- **Accelerate change and address educational barriers** — ASEAN students can find their place in the Industry 4.0 economy, but they will need help. The Industry 4.0-related challenges of each ASEAN member country vary widely. Each nation's stage of economic development, economic structure, and growth policy priorities are unique. These factors affect the work-related knowledge and skills needed to adopt advanced technologies.

- **Develop new ways of teaching for the modern knowledge economy** — The current educational approach used in ASEAN classrooms doesn't equip students for the Industry 4.0 knowledge economy. School-centred education emphasises memorisation, facts, and a top-down approach to learning.

186

Whether students receive training for the professions or the factory, they are more likely to succeed when they're equipped with problem-solving and critical-thinking skills.

- **Help students learn how to learn** — This ability will be an essential part of education throughout a student's life. ASEAN educators are already equipping students with this highly valuable skill. However, throughout the region, more effort is needed to help students develop the habit of lifelong learning.

- **Develop new teaching methods for the "multi-career" age**. In many developed countries, single-employer careers are a thing of the past. Single-career lives spent in one knowledge domain are waning, too. It's not unusual for professionals to work for a series of employers in one knowledge domain. Or, employees can move from one related professional speciality to another as job availability changes.

In this an uncertain unpredictable work environment, but life-long learning can help employees boost their employability and make transitions to new careers or knowledge domains.

Barriers that ASEAN educators can overcome — Consistent educational policies and clear goals are an essential part of the region's educational transformation. Unfortunately, ASEAN countries do not share a clearly defined education policy. Each country prioritises and focuses on different aspects of its educational system. According to McKinsey, ASEAN's economic growth trajectory predicts that the region will become the world's fourth-largest economy by 2030. However, action—not in evidence in many ASEAN nations so far—is needed to develop human capital and workforce skills in manufacturing.

Educators in each ASEAN nation face unique difficulties in developing work-related skills and knowledge. Here are some examples of the status of education in ASEAN-6 countries with a manufacturing base:

187

- Singapore has always maintained a strong connection between schools and industry in Singapore. However, there's a need to reform curricula that use a hierarchical, top-down approach to education. Now, education will enable agile thinkers, who can change along with Singapore's innovation-driven economy. Singapore's educational system will also nurture an innovation mind-set. This new way of thinking combines curiosity, creative thinking, the ability to collaborate, and the willingness to take risks.
- Malaysia continuously strengthens its semiconductor, solar power, and light-emitting diode industries. However, the country needs further investments in human capital development to have an adequate pool of skilled talent for high value-added activities. This goal requires Malaysian manufacturers to act. They must fill the gap between the knowledge domains of its graduates and the skills that its industries need.
- Thailand's is developing manufacturing production based on the latest knowledge and technology-intensive methods. However, a shortage of workers with technical backgrounds hampers this effort. The country's education system calls for training more workers with technical skills and more students with science and engineering degrees.
- Indonesia does not have enough young students with the skills that would help them become part of the digital manufacturing economy. This is especially true in technical, scientific, and engineering fields.
- Industry 4.0 adoption requires aggressive reforms to the public education system. Stronger relationships with educators and industry are especially important. Indonesian education would also benefit from providing more continuous

learning opportunities and more formal training in entrepreneurship.

- In the Philippines, there's a yawning gap between the nation's schools and industries, which challenges educators wanting to create opportunities enabled by Industry 4.0. The likelihood of integrating Industry 4.0 technology and methods into the nation's economy is uncertain. Unless, of course, national officials have the political will to allocate more resources to education reform and technical training.

11.4 ASEAN Manufacturers

Industry 4.0 technologies offer ASEAN manufacturers a golden opportunity: the chance to capture a larger share of global industrial markets. First, they must bridge the productivity gap. The days of competing with low-cost workers alone are gone.

The desire of ASEAN manufacturers to compete more effectively in world markets is genuine. However, the initial results of Industry 4.0 integration efforts are uneven. There's little or no region-wide strategy or view of the future. It's up to each manufacturer's executives to monitor and guide the development and execution of their own Industry 4.0 roadmaps.

11.4.1 Transforming Knowledge into Action

Yes, it is every company for itself. However, many ASEAN manufacturers know that digital design and process control technologies can transform their operations and improve their bottom line. They know that improving the region's relatively low manufacturing productivity is an important part of digital transformation.

However, ASEAN manufacturers have only begun exploring the possibilities that Industry 4.0 technologies make possible. For

189

example, only 30 to 40 per cent of production lines in the region are automated. The remaining lines use discontinuous, labour-intensive processes that frequently create quality assurance problems.

11.4.2 Digital Supply Chains

One of the most significant elements of Industry 4.0 is the extended view of manufacturing and production. Once a matter of operations processes, manufacturing now embraces design, sourcing, production, distribution, and post-sales activities.

Throughout the world, operations processes are becoming more connected, digitised, transparent, and efficient. This is a good thing for modern digital supply chains, which rely on the availability of:

- Integrated project planning and management systems.
- Autonomous logistics and smart procurement software.
- IT advances such as cloud computing, big data analytics, and internet-connected devices on the Industrial Internet of Things (IIoT).

More efficient digital supply chains can be especially helpful in ASEAN nations, where operations can be inefficient. This makes delivering goods when and where they're needed can be challenging.

Technology, which will help supply chains to become more efficient, is a key driver of more effective manufacturing operations. From ethical and business points of view, manufacturers must provide the ultimate in supply chain traceability and security. So, it makes sense to use the latest advances in process control and security, such as Blockchain technology.

11.5 Who's ready for Industry 4.0?

Experience tells us that not all manufacturers are equally ready to benefit from Industry 4.0 developments, but which companies are ready to adopt Industry 4.0 technologies?

Industry 4.0 readiness is a complex idea, which we discussed at length in Chapter 8. In the context of the status of Industry 4.0 adoption, a more straightforward, indicator-based approach will suffice. Below is a short profile of companies that are most likely to benefit from implementing Industry 4.0 technologies and ideas:

- Large companies
- Digitally mature companies
- Those engaging Industry 4.0 technologies in several use cases
- Those digitised their core businesses.
- Businesses that clearly define target values by critical success factors.
- Businesses whose company executives support Industry 4.0 adoption efforts.

11.5.1 Early Adoption Plans

Some companies in ASEAN have already started to introduce elements of Industry 4.0 technologies into their business models.

For example, operations of major semiconductor manufacturer, Infineon Technologies, plans to invest more than $84 million in a smart factory in Singapore. The company expects improved operations to cut production cycle times in half, increase productivity by 10 per cent, and save $1 million a year in energy costs.

Start-ups focused on Industry 4.0 technologies are also emerging in the region. Singapore-based KaHa, a smart IoT wearables manufacturer, is attracting funding from investors. Other start-ups such as Kuala Lumpur-based N'osairis have attracted millions in development funds from domestic and foreign investors.

11.5.2 Accelerators & Barriers

- Industry 4.0 technologies are accelerating manufacturing time to market and boosting resource and labour productivity. Although ASEAN manufacturers have been slow to start Industry 4.0 adoption, they can achieve long-term value by setting up digital transformation initiatives. Even though it's early days, these trends have accelerated Industry 4.0 adoption:

- **A dramatic increase in data volume, computing power, and connectivity** — Now, ASEAN manufacturers have more information to use and improving support technologies such as ICT.

- **Development of advanced sensor and control devices** — These internet-connected devices enable new product delivery models and add value to familiar products, such as highly customised smart apparel.

- **Development of big data analytics and business intelligence capabilities** — Helps manufacturers identify ideal maintenance frequency and changes in customer preferences for clothing products.

- **Successful commercialisation of advanced IT** — such as virtual and augmented reality devices and cloud-based service delivery.

- **Advanced production methods** — 3D printing and other emerging production technologies enable new approaches to process and product design.

- **New forms of human-machine interaction** — Digitised and automated processes now use new interface options.

192

Augmented reality, virtual reality, and advanced robotics enable workers to operate alongside collaborative robots known as cobots. These machines handle physically demanding and repetitive tasks, and their programming now makes them safe for humans to work beside.

ASEAN manufacturers can speed up the region-wide adoption of Industry 4.0 technologies by joining other manufacturers to test ideas. Throughout ASEAN-6 nations, companies tend to collaborate within their own organisations. However, technology adoption could be accelerated if companies within ASEAN collaborated with each other to test ideas.

Forming regional and global partnerships would be a good way to do this. In addition to local alliances, ASEAN companies should consider building deep relationships with global manufacturers and technology providers to ensure access to the latest technologies.

Obstacles to digital manufacturing development — There are plenty of barriers that slow manufacturers' adoption of Industry 4.0 technology. Within companies, these problems include:

- Inability of companies to redesign their business plan for an Industry 4.0 world.
- Persistent data silos.
- Lack of technical talent with the skills and knowledge needed to create and implement an Industry 4.0 roadmap.
- Old-fashioned risk aversion — Many companies have concerns about the unknowns of Industry 4.0 adoption. For example, an uncertain cybersecurity environment and high fixed capital costs make taking the risk of Industry 4.0 adoption unappealing.
- A workforce out of sync with market demand — Technology drives the need for technically skilled workers, who are often difficult to find.

ASEAN manufacturers struggle to set priorities based on the potential impact of Industry 4.0 use cases. Their difficulties arise from a lack of experience with available technology and a shortage of data makes economic assessments a daunting proposition.

11.5.3 How ASEAN Nations are Organising

We've described the roles of Industry 4.0 ecosystem participants and what they can do to accelerate the adoption of digital manufacturing, but what about their interactions, the relationships that tech adopters and stakeholders must engage in?

ASEAN government officials, academics, and manufacturers must organise to get region-wide adoption off the ground through a full spectrum **collaboration**.

Regional economic integration is not a new idea among Southeast Asian nations. Since 2015, the ASEAN Economic Community (AEC) has been designed to minimise trade barriers and establish a common market. Eventually, the AEC would enable unrestricted flows of goods, capital and labour throughout the region. Although member countries have very different economic and political systems, the AEC framework could become a key component in the Industry 4.0 adoption process.

Forward-looking supporters of the AEC and Industry 4.0 point to several positive outcomes of technology adoption within ASEAN. These include better awareness of Industry 4.0 technologies and the need to adopt them. Also, regional activity can encourage the use of a wider variety of enabling technologies. Use of semi-independent robots and IIoT devices are familiar topics to many ASEAN manufacturers. However, use cases of less-familiar applications such as 3D printing gain immediacy with applications closer to home.

12   Current ASEAN Use Cases

We've looked at the Industry 4.0 framework and the value that ASEAN countries expect to achieve, but how are major ASEAN manufacturing countries doing in their quest to adopt Industry 4.0 technologies?

It's no secret that different ASEAN nations have made varying degrees of progress and face different challenges. So, we'll profile six ASEAN countries with established manufacturing capabilities, describing how each develops its digital manufacturing capabilities.

12.1 Singapore: Plans for Manufacturing Future

Although manufacturing has lost some of its lustre in Singapore, the country's vision is to revitalise its manufacturing sector and become a hub in global advanced manufacturing. Singapore's manufacturers recognise their growing operations costs, and future domestic labour shortage. They realise the need to modernise their manufacturing model to one that delivers more innovative, high-value products. However, manufacturers also expect Industry 4.0 technologies to deliver simpler, more flexible production processes and faster turnaround times.

The government has repeatedly declared that manufacturing will remain a cornerstone of Singapore's economy. Here are the ways that Singapore uses the Industry 4.0 adoption process to achieve this goal.

- Singapore started with the Republic's Smart Cities initiative.
- Key sectors have rolled out Industry Transformation Maps.
- Companies now use the Singapore Smart Industry Readiness Index (SSIRI).
- The government launched Standards Mapping for Singapore Smart Industry Readiness Index in March 2018; the mapping document enables businesses to identify the standards that

apply to establishing Industry 4.0 capabilities. A companion document to the SSIRI,

- Adoption of Standard, In 2019 Singapore Standards Council (SSC) has adopted IEC 62541 (OPC UA) for communication and data exchange between machines, systems and infrastructure in digital factory as national standard.

12.1.1 Singapore's in the World of Ind. 4.0

In the World Economic Forum's 2018 Global Competitiveness Report, Singapore was identified among 25 "leaders in manufacturing today that are also well positioned for the future of production." Singapore ranked second in the drivers of production category, which describe key enablers that help a country take advantage of advanced manufacturing capabilities. In this category, Singapore ranked second only to the United States.

Singapore ranked 11th in the "structure of production" category. This metric measures the size and complexity of a country's manufacturing sector. Other countries in the top 25 include Japan, South Korea, Germany, and the United States. Leading countries in this group have a first-mover's advantage. The Republic has many advantages, which could result in Industry 4.0 benefits. These include:

- Extensive manufacturing capabilities in high-value areas such as product research, development, and design.
- Strong collaboration among government officials, business owners, and trade unions. These stakeholders participate in extensive consultations on how to move the city-state's economy up the global production value chain.
- The beginnings of SME adoption of Industry 4.0 technologies. Disruptive technologies such as 3D printing, IoT, and robotics. For example, Osteopore International Pte uses 3D printing to

customise bone or skull implants built with bioresorbable polymers.

Throughout the Singaporean manufacturing sector, business owners know that they can't rely on low-wage, high-volume manufacturing in the 21st century. Macroeconomic factors such as a shrinking labour force are drawing their attention to the need for thoroughly modernised operations. Workers—professionals and machine operators—must gain new knowledge and master new skills needed for jobs throughout the product life cycle.

These new requirements and the wish to compete via innovative technology-based solutions make the country's manufacturers keen to adopt Industry 4.0-based solutions.

No ASEAN country has a perfect Industry 4.0 readiness record. However, Singapore leads its regional neighbours in this regard. Perhaps that's because many incentives, policies, and support programs are available to smooth the bureaucratic and technological paths to Industry 4.0 adoption. Here are a few of the many ways that the Singaporean government support Industry 4.0 adoption:

• **Government-sponsored trade shows** — The Singapore government has been working with multinational companies for a very long time. For instance, the government hosted Industrial Transformation Asia-Pacific (ITAP) in October 2018 and 2019. This trade expo featured the latest manufacturing technologies from German and other multinational corporations.

The ITAP conference enabled attendees to move from the awareness stage (recognising the importance of Industry 4.0 transformations) to implementing Industry 4.0 technologies in their own factories. Through ITAP demonstrations and lectures, manufacturers experienced real equipment engaged in real-world applications of the latest technologies and solutions.
The emphasis was on solutions and technologies that companies could adopt immediately. Event organisers made sure to present

topics and equipment that were practical and applicable to a Southeast Asian audience. By using this educational slant, ITAP informed the public about how manufacturing is evolving and how this can ultimately help Singapore's workforce get better jobs and career opportunities.

• **Government-business partnerships** — In Singapore, there are many ways for industry players and manufacturers to collaborate, share resources, and achieve technology breakthroughs. Take the Advanced Remanufacturing Technology Centre (ARTC), for example.

The centre is led by The Singapore Agency for Science, Technology and Research (A*STAR) in partnership with Nanyang Technological University. Industrial technology specialists from more than 60 multinational corporations work with Singaporean partners to develop advanced manufacturing and remanufacturing capabilities. For example, Swiss industrial technology giant, ABB, is a partner. At the centre, ABB technologists work closely with those from NTU, A*STAR, and other multinational corporations to develop remanufacturing applications that cost less and use materials more efficiently.

• **Government-sponsored model factory** — The ARTC also started a model factory. Manufacturers can use its live production environment to learn, experience, and experiment with advanced manufacturing technologies.

The facility includes a virtual manufacturing lab, which uses virtual and augmented technologies. Using these advanced technologies, manufacturers can try virtual factories or machine operations before they build or lease the real thing. The virtual and augmented reality capabilities also enable training and skills development for workers.

• **Worker training efforts** — To replace or improve worker skills, the government works closely with trade unions. Trade union leaders play an essential role in encouraging workers to participate in training for new jobs. Some basic course training is free. Other courses are co-funded by the government and firms that send their employees to be trained.

• **Rethinking technical education** — Industry 4.0 manufacturing environments require workers with up-to-date technical knowledge and skills. Satisfying these requirements demands education with a good foundation in science, technology, engineering, and mathematics (STEM). Singapore's problem: a significant part of the country's workforce is neither good at nor interested in STEM subjects.

Digital transformation expectations are high for many firms in Singapore. However, successful Industry 4.0 evolution requires something that goes beyond knowledge of technology. Success will require a new type of education and a new sense of business value.

• **Getting Singaporeans educated** — Industry leaders in the Lion City recognise that benefiting from Industry 4.0 technologies requires employees to master new, more relevant, technical knowledge and skills. Government, business, and academic leaders in Singapore also need to rethink how to develop citizens for work, especially technology-related work.

The need lies not just in teaching STEM subjects but making sure that employees develop a mind-set that develops a sense of growth and value. Not all Singaporean employers see the need to invest in human capital development. Some do not view well-trained employees as a competitive advantage. A new approach to education and training requires a fundamental investment in soft skills such as strategic thinking, problem solving and collaboration.

12.2 Malaysia

Of the five largest manufacturing countries of ASEAN, Malaysia is the second wealthiest in terms of per-capita GDP. Its Industry 4.0 status includes established frameworks and policies and the beginnings of technology implementation by early adopters and the support of the national government.

12.2.1 Industry 4.0 in Malaysia, Today & Tomorrow

Government and business leaders have noticed how application of information technology and the Internet of Things has opened market supply chains. This eases access of products from the most remote and rural areas, sending economic prosperity the other direction.

12.2.2 Possible Slowdown in Adoption

Early adopters have taken the plunge and started using digitised technologies in their factories. However, many other manufacturers, especially SMEs, aren't sure that now's the time to get onto the Industry 4.0 bandwagon. This reluctance is just one of several factors that could slow the pace of adoption in Malaysia.

A slowdown of digital technology adoption would undermine the government's ambitious vision of Industry 4.0-based growth and prosperity. Here are the program targets:

- A 30 per cent, per capita increase in manufacturing productivity.
- An increase in the absolute contribution of the manufacturing sector to the economy from RM254 billion to RM392 billion.
- Strengthen innovation capacity and capability as reflected by improvement in the Global Innovation Index ranking from 35th to the top 30.
- Increase the number of highly skilled workers in the manufacturing sector from 18 to 35 per cent of Malaysia's workforce.

These targets are part of a program of Malaysia's Industry 4.0 strategies, policies, and incentives.

12.2.3 Malaysia's Industry 4.0 Program

Malaysian government officials look to Industry 4.0 as an enabler of all the usual capabilities: enhancing global competitiveness, strengthening the growth of the manufacturing sector, reducing dependence on foreign labour, and so on. To realise these goals, representatives of Malaysian government ministries, businesses, and academe collaborate to establish a framework that guides Industry 4.0 adoption.

The government's role — On 31 Oct 2018, Tun Dr. Mahathir Mohamad, Malaysia's Prime Minister launched Industry4WRD, the nation's Industry 4.0 policy document. MITI, which has worked on the policy since mid-2017, said the government will act as an enabler in five areas — funding, infrastructure, regulatory frameworks, skill improvement programmes, and technology accessibility. The goal: to reach the specific targets described above by 2025.

Government policies focus on five manufacturing industries: electrical and electronic devices, machinery and equipment, chemicals, aerospace, and medical devices. Malaysia's 2019 budget is an encouraging indicator of support. The new budget includes RM2 billion to encourage SME adoption of emerging technologies. The government also subsidises the Industry Digitalisation Transformation Fund with an additional RM3 billion investment. The goal is to drive the adoption of smart technologies such as AI and those used in automated manufacturing.

The national government also encourages companies, especially SMEs, to use these facilities and programs to advance Industry 4.0 adoption. Here's a partial list of government support programs:

- Domestic Investment Strategic Fund provided by MIDA

- The MIDA-sponsored Automation Capital Allowance (ACA)
- Soft Loan Scheme for Automation and Modernisation
- Accelerated Capital Allowance and Capital Allowance for purchase of ICT equipment and computer software packages
- Talent development and training opportunities provided at the German-Malaysia Institute and other skills development centres.
- A national regulatory sandbox initiative. — A model program that encourages innovators to test their ideas in a real production environment.

Industry 4.0 adoption in manufacturing. Datuk Azman Mahmud, the CEO of MIDA, describes current automation capabilities, saying, "[As of July 2018], there are about 120 major companies in Malaysia that can produce advanced handling systems for full automation and incorporate intelligent robots, including machine-to-machine communication."

These early adopters are not typical of Malaysian manufacturers. However, the country's manufacturers are aware of information technology, such as the Internet of Things (IoT). They also understand how digital technology has opened market supply chains and made access to products from the most remote areas easy.

Automation offers Malaysian manufacturers the benefits of higher levels of product quality and consistency, more efficient operations, greater production output, energy savings, fewer human errors, and often a faster return on investment.

Automation in the plastics industry — A good example of current automation is in plastics manufacturing, where most basic processes have been automated by installing robotic arms and conveyor belts. More advanced automation, such as automated visual inspection, auto-reject detection, and auto packing, requires expensive investments in technology and equipment. These processes are automated less often.

Mr Azman provides another example, saying, "United Sweethearts Garment, increased its production volume by more than 300%... By automating its operations, the company has also managed to reduce its defect rate by about 85 per cent."

The company plans to gradually reduce its foreign workforce over the next 5 to 10 years as it adopts more and more automation. The company's short-term goal is not to reduce labour but to use the same number of workers to produce more goods. Analysts predict that more Malaysian SMEs will adopt automation and other Industry 4.0 technologies in the next few years.

12.2.4 What Drives Ind. 4.0 Adoption Forward?

The availability of automation technology drives Industry 4.0 adoption to the point that many manufacturers view automation as inevitable. However, rising labour costs and the nation's desire to improve the country's productivity also moves Industry 4.0 adoption forward.

The Malaysian Prime Minister, Tun Dr. Mahathir Mohamad, states what many manufacturers already know — improving manufacturing productivity is the crux of improving Malaysian prosperity. They also recognise that manufacturers can no longer depend on cheap labour and plenty of capital to stimulate productivity. Industry 4.0 is Malaysia's alternative route to prosperity and competitive success with its ASEAN neighbours.

12.2.5 Gaps in Infrastructure, Skills, & Technology

Although Malaysia has made impressive progress in adding digital technology to its infrastructure, gaps in its telecom technologies remain. The country's industrial areas still need more high-speed internet coverage to support processes such as visual automation.

For example, Telekom Malaysia will have to invest in fibre-optic cables that can deliver 100 Mbps Unifi, high-speed internet communications. There are other potential obstacles to adopting Industry 4.0 technologies and practices:

Automation technology — SMEs involved in low- to mid-volume manufacturing recognise that adopting flexible automation technology and processes is their big challenge. However, they often face the high cost of equipment and a lack of understanding about automation technologies. For example, a common misconception is that adopting automation technology is always expensive and that manufacturers must automate their entire production line at one time.

Finding skilled help — Many manufacturers need technical help in getting started with automated equipment and processes. Malaysia could use more automation consultancies. Some manufacturers report that they work with foreign consultancies, which deliver world-class services but charge high fees. Others would work with local or regional consultants, if they were available.

Bureaucratic matters — Manufacturers find it difficult to handle some of the paperwork and documentation for the Malaysian government. The challenge: understanding whether they qualify for various programs. They need help in getting the application process started. Manufacturers would like government agencies to reduce the red tape in the application process and provide clearer guidelines.

12.3 Thailand

Thailand has made a definite but uneven start in its Industry 4.0 adoption process. Frameworks, policies, and incentives are set, and early adopters use robots and automation at the process level. However, Thai manufacturers still haven't extended Industry 4.0 use along the production hierarchy or to other parts of the product value chain: R&D, product design, logistics, and marketing.

204

It's still early days for Industry 4.0 in Thailand. The government started on the right foot by establishing goals, high-priority industries, and incentives for Industry 4.0 adoption. In October 2018, the country introduced Thailand 4.0, its Industry 4.0 initiative. The program is a result of the Thai government's vision of a new economic model. Its goal: to help the Thai economy avoids the dreaded middle-income trap.

The newly approved Industry 4.0 plan gives top priority to 10 key industries. Robotics and logistics get the top spots and a $6 billion share of the $45 billion program budget. The Thailand 4.0 strategy emphasises using advanced technology to improve process efficiency on the factory floor. However, it also includes the development of growth hubs, technology clusters, and start-up-based robotics, IoT, and biotechnology.

12.3.1 Driven by Opportunity & Competition

Thailand is still an industrial powerhouse in Asia ($400 billion GDP in 2017). In recent years, however, the country has witnessed an unspectacular, average 3.7 per cent GDP growth rate. Moreover, Thailand is on the cusp of the dreaded middle-income trap, where wage inflation makes industries less competitive. Thai government officials hope that Industry 4.0 technologies will help manufacturers boost their process and worker productivity and revitalise their competitive standing in ASEAN.

Thailand has gone to extraordinary lengths to create tax breaks and incentives that support automation and robotics. These plans and programs include corporate tax exemptions (up to 13 years) for businesses that use advanced technology or engage in Industry 4.0 related R&D activities. Here are some examples:

EEC Project Inducements — Incentives for Thailand's Eastern Economic Corridor (EEC) project encourages businesses to add robots and robotics technologies to their manufacturing and production processes.

Serious money — In February 2018, the Thai government authorised $45 billion to finance the country's leap into a technology-driven economy. More than $6 billion is slated for robotics and logistics technology upgrades.

Plenty of support and incentives — Thailand has invested in many support programs and networks. Thai educational institutions play a role in supporting research and development. For example, the Institute of Field Robotics (FIBO) of King Mongkut University of Technology in Thonburi currently offers undergraduate and graduate programmes in robotics and automation engineering.

Thailand has gone to extraordinary lengths to create tax breaks and incentives that support automation and robotics. These plans and programs include corporate tax exemptions (up to 13 years) for businesses that use advanced technology or engage in Industry 4.0related R&D activities. Here are some examples:

12.3.2 Barriers to Ind. 4.0 Progress in Thailand

Progress during the next three years will show whether Thailand is transforming itself into a technology-driven economy. Thai manufacturers will need large helpings of education reform, foreign investment, and technical help to prove that the answer is "yes."

The need for a high-tech workforce — Thailand's ageing population and a relatively unskilled workforce drive the need to invest in a modern, Industry 4.0-ready workforce. Thai officials and educators will have to work very hard indeed to create a home-grown cadre of data scientists and analysts, engineers, and technology specialists. There's also more work in store for those who will educate and train skilled operators and maintenance support workers, who will run and repair automated equipment.

A need for more foreign direct investment — If Thailand will upgrade its economy, the country needs another infusion of foreign direct investment. Achieving this goal will require new efforts to link multinationals with domestic innovators and local manufacturers.

The need for more automation consultants — These indispensable professionals provide the knowledge, experience, and guidance that manufacturers need when they set up their newly automated manufacturing processes. Currently, about 200 automation consultants work in Thailand. The government wants to see 1,400 consultants in residence by 2023.

12.4 Indonesia

Indonesia, like several of its ASEAN neighbours, has a proud history as a vital, manufacturing-based economy. Like its neighbours, Indonesia must add new methods and business models to its toolkit to compete successfully in an Industry 4.0 environment.

The Indonesian government has high hopes for its Industry 4.0 manufacturing growth. Targets include:

- A rise in the nation's net export-to-GDP ratio from 1 per cent in 2016 to 10 per cent by 2030.
- Creation of 7 million to 19 million jobs, which will contribute a sustainable 1 to 2 per cent to GDP growth by 2030.
- Growth of the manufacturing contribution to the national GDP from 18 per cent in 2016 to 21 to 26 per cent by 2030.
- Increases in employee productivity, global export market share, and job opportunities, as well as a return to its former status as a leader in net exports.

These are high hopes, indeed. However, Indonesia has laid a foundation of these goals by creating its own Industry 4.0 framework, policies, and incentives.

12.4.1 Frameworks & Policies

Indonesian policymakers think that disruptive Industry 4.0 technologies are just the ticket to help make the nation's manufacturing sector more efficient and more competitive in the region and beyond.

To start the Industry 4.0 adoption process, the Indonesian government officially launched the nation's official roadmap, Making Indonesia 4.0, in April 2018. The initiative framework named five high-priority technologies: hi-speed internet, artificial intelligence, human-machine interface, 3D printing, and robot and sensor technology to boost industrial capacity and rapidly increase production output.

The framework was supported by a partnership with the Singaporean government. In November 2018, partners Enterprise Singapore and the Indonesian Agency for Research and Development of Industry (ARDI) signed an agreement to develop and adopt Industry 4.0 technologies by Indonesian manufacturers.

Driving manufacturing back to its former prominence — Now, as in the 2000s, manufacturing plays a vital role in Indonesia's economy, employing more than 14 million people. However, the sector has been in decline for some time. As services become an increasing part of the national economy, manufacturing's contribution to GDP has declined from its peak of 29 per cent in 2001 to 20 per cent in 2018.
Though challenging, the opportunity to replace China as the factory of the world is very appealing. The major obstacle to this goal, however, is production and labour productivity. Like many ASEAN countries, Indonesia cannot prosper as a low-wage manufacturer. To stay competitive and stay out of the middle-income trap, the nation must resort to disruptive technologies and new business opportunities. The

desired goal: use new Industry 4.0 technologies to boost productivity and enable Indonesia to produce higher-value goods.

Sweetening the pot with incentives — Indonesia offers a standard package of incentives, such regulations and policies aligned with Industry 4.0 goals. Tax rebates and subsidies may be offered to speed up the implementation of selected Industry 4.0 technologies, for example. The Making Indonesia 4.0 program supports greater collaboration between representatives of the government, private sector, and universities to develop an innovation-based network of public and private sector partners.

Playing catch-up and resistance to change — Like several of its ASEAN neighbours, Indonesia faces severe shortages of modern manufacturing technologies, communications infrastructure, and workers skilled and ready for work in an Industry 4.0 world.

The success of Industry 4.0 modernisation will depend on Indonesia's ability to modernise these essential resources and integrate them into their legacy manufacturing operations:

- Outdated secondary and university education — Building a skilled technical workforce is one of Indonesia's biggest challenges. Current initiatives are designed to modernise the nation's secondary and university educational systems and curricula to hasten technical learning and an understanding of STEM topics.
- Not enough technical consultants and technology specialists — Currently, Indonesia depend on foreign workers to provide consultancy and specialised technical knowledge, skills, and experience. Gradually, the demand for foreign specialists is expected to drop as the country develops its local workforce. For now, the need for local (and more affordable) consultants is growing.
- Modernising the business infrastructure — Bringing Indonesian education into the 21st century isn't its only Industry 4.0 challenge. The country is still in the process of

buying and integrating connectivity, information, and communications technologies into its business and manufacturing infrastructures. Without widespread connectivity, Indonesia's Industry 4.0 effort will be limited to model factories and innovation hubs in established industrial areas.

- Modernising the business mind-set. Perhaps the most challenging obstacle to Industry 4.0 adoption is the inherent conservatism of many Indonesian manufacturers. In a recent survey, 52 per cent of potential adopters indicated interest in and awareness of Industry 4.0 technologies. However, only 13 per cent of these respondents have committed to implementing Industry 4.0 technologies.

Security and risk concerns — Other surveys show Indonesian manufacturers are still concerned about cybersecurity or financial risk caused by what they see as high capital costs. Perhaps the most worrying indicator is that manufacturers aren't aware—or don't believe—that there's plenty of value to be an early Industry 4.0 adopter. Companies might be waiting for the price and uncertainty of specific technologies to decline before they risk an investment.

12.5 Vietnam

In Vietnam, when you read or hear descriptions of Industry 4.0 progress, the words "research," "studying," and "determination" frequently appear in government announcements and articles in the press. A range of companies, large and small, are using emerging technologies such as AI and Blockchain. However, Vietnam is still in the early stages of Industry 4.0 adoption.

12.5.1 High Hopes & Slow Starts

According to a survey conducted in late 2018 by the Ministry of Industry and Trade, 82 per cent of Vietnamese businesses were not prepared for the fourth industrial revolution. Only 10 per cent of respondents said that they were ready.

Now, Vingroup, the nation's biggest corporation, recently announced plans to invest in AI, big data analytics, and advanced software development. The company also plans to create a high-tech industrial park to encourage collaboration between entrepreneurs and engineers, who will develop data analytics and other IT applications. Vingroup believes that this move will drive Vietnam's high-tech industry onto the world stage.

Vietnam's government officials hope to enhance technological and manufacturing capabilities throughout the global value chain. They're looking forward to creating new and sustainable business models for start-ups and dramatically improving manufacturing competitiveness in the future.

The Vietnamese government is determined to create focused, long-term Industry 4.0 policies. The Prime Minister's Directive No. 16/CT-TTg, "On the strengthening of the ability to access the Fourth Industrial Revolution" (May 2017) provides guidance for the country's early-stage Industry 4.0 adoption efforts.

The government identified five high-priority technologies for the nation's Industry 4.0 initiatives. These technologies include the Internet of Things, artificial intelligence, human-machine interfaces, robotics and sensor technology, and 3D printing. However, as of May 2019, the country's Industry 4.0 strategy document, "National Strategy on the Fourth Industrial Revolution," is still being developed. Current economic and trade trends — Vietnam's robust economic growth is driven by vigorous manufacturing and export expansion. In addition to rising domestic consumption, strong investment fuelled by FDI, and vibrant domestic enterprises. Vietnamese manufacturing has benefitted from a robust foreign direct investment employed about 2.4 million workers.

Now, Cambodia and Myanmar are competing with Vietnam for low-skilled production jobs. Foreign trade is slowing, which makes competition for customers fiercer than ever. Vietnam has been able to mobilise an abundant supply of young, well-educated workers. Throughout the past decade, this rich labour attracted foreign direct investment in labour-intensive manufacturing.

Times have changed, however, now Vietnamese manufacturers look forward to using modern technology to satisfy the increasing demand for manufactured goods from an expanding consumer class in nearby Asian markets.

Overcoming major obstacles — In the World Economic Forum's Readiness for the Future of Production 2018 report, Vietnam scores behind Singapore, Malaysia, Indonesia, and Thailand in the ASEAN-6 group. This position reflects two major problems that block Vietnamese manufacturers from benefiting from disruptive digital technologies.

Stuck in the technological past — There's a reason why Vietnam's Industry 4.0 readiness score is only 18 per cent of surveyed companies. Manufacturers still use machines, devices, tools, and methods from earlier industrial revolutions. This lack of readiness is due largely to unequal industrial development in different regions of Vietnam. Manufacturers won't be able to adopt Industry 4.0 technologies until they modernise their communications and connectivity infrastructures.

The qualifications-requirements gap — There's a gap between the qualifications of Vietnamese workers and the needs of manufacturers. The gap is large, and it's growing. If not addressed, this problem will put a damper on Vietnam's development aspirations.

Progress is being made in strengthening vocational training and higher education. However, Vietnam needs to invest more heavily in these areas. The government and universities should work more closely together to produce enough qualified professionals, machine operators, and maintenance workers to staff Industry 4.0 facilities.

12.5.2 Vietnam's Current Industry 4.0 Status

So how can we describe Vietnam's current status as a nation of Industry 4.0 practitioners?

In 2018, Vietnam launched the nation's "National Action Plan for Implementing the 2030 Programme for Sustainable Development." The announcement repeated that science and technology are the foundations of the country's sustainable economic development.

The government's recent development efforts have been consistent with this view. However, creating and implementing Industry 4.0 policy remain very slow processes. Vietnam's Industry 4.0 to-do list is long, and the nation's ASEAN competitors are moving ahead apace. Here's a brief list of what still needs to be done:

Government policy changes. National-level changes should include:
- Providing adequate resources and finances to invest in research facilities and create an environment where innovation is encouraged.
- Creating clear policies and a strategy that would direct education and practice of science and manufacturing-related technology and engineering

Education and training measures should focus on:
- Improving the quality of domestic scientific and engineering education.
- Stimulating new educational and training methods as part of business development.

Vietnam's businesses can focus on practices that:

- Take advantage of international academic and multilateral development organisations to transfer advanced knowledge to local research institutions.
- Focus more on market-related inducements to digital manufacturing and less on the push to use specific technologies.
- Promote the habit of collaboration between Vietnam's business, government, and educational bodies. Include multilateral organisations, such as the Asian Development Bank, World Bank, European Union, the United Nations, and multinational corporations.

12.5.3 Global & ASEAN Industry 4.0 Status

Since 2011, countries announced and launched Industry 4.0 or similar initiatives, starting with Germany and later the United States and Japan. The ASEAN community took its cue to follow suit with various programs and initiatives. Figure 12-1: The Global and ASEAN Implementation of Industry 4.0, provides a picture of the Industry 4.0 implementation landscape.

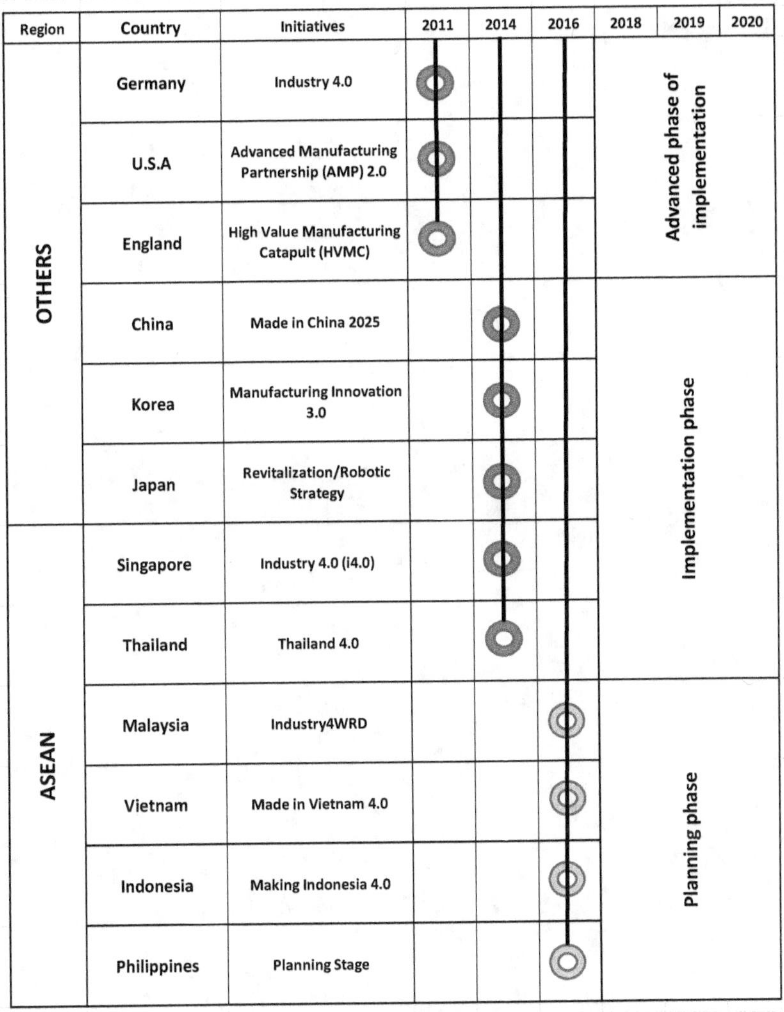

Region	Country	Initiatives	2011	2014	2016	2018	2019	2020
OTHERS	Germany	Industry 4.0	◎				Advanced phase of implementation	
OTHERS	U.S.A	Advanced Manufacturing Partnership (AMP) 2.0	◎				Advanced phase of implementation	
OTHERS	England	High Value Manufacturing Catapult (HVMC)	◯				Advanced phase of implementation	
OTHERS	China	Made in China 2025		◎			Implementation phase	
OTHERS	Korea	Manufacturing Innovation 3.0		◎			Implementation phase	
OTHERS	Japan	Revitalization/Robotic Strategy		◎			Implementation phase	
ASEAN	Singapore	Industry 4.0 (i4.0)		◎			Implementation phase	
ASEAN	Thailand	Thailand 4.0		◯			Implementation phase	
ASEAN	Malaysia	Industry4WRD			◎		Planning phase	
ASEAN	Vietnam	Made in Vietnam 4.0			◎		Planning phase	
ASEAN	Indonesia	Making Indonesia 4.0			◎		Planning phase	
ASEAN	Philippines	Planning Stage			◎		Planning phase	

Figure 12-1: Global and ASEAN Implementation of Ind. 4.0

13 Resources: Lifeblood of Digital Transformation

"Human potentialities constitute the world's greatest resource."

Julian Huxley

Resources are the fuel that makes digital transformation happens. In this chapter, we're focusing on the enabling resources of ASEAN member states, including:

- Converging information and operations technologies, which enable the cyber-physical systems of smart manufacturing.
- Human resource development, which includes improved education, training, and skills development.
- Connectivity infrastructures such as fibre-based and fixed-bandwidth services and narrow-band IoT.
- Private and public-sector funding programs, which provide loans, grants, and direct foreign investment for digital manufacturing projects.

This chapter presents country profiles, which describe each ASEAN-6 nation in terms of the Industry 4.0 development assets that are available today.

In "What Singapore's Smart Industry Index Can Tell Us About Industry 4.0 in 2018," author Charing Kam describes a critical aspect of the Singapore Smart Industry Readiness Index. She encourages manufacturers to think beyond their current systems and processes and move beyond the mere provision of products and equipment.

Manufacturers throughout ASEAN, she says, will have to make significant changes to become or stay competitive. Government, academic, and business entities will have to throw funding and new ideas into the pot of Industry 4.0 innovation.

Manufacturers in ASEAN-6 countries have a wide variety of resources to use in their digital manufacturing development strategies. The types of resources available to each country leave their unique mark on its readiness now and on its national GDP in the future.

Here are detailed summaries of resources that each ASEAN-6 nation can use to make the most of their Industry 4.0 development opportunities.

13.1 Singapore

The consensus of analysts is that Singapore's supporting policy, advanced technology, and robust business infrastructures put the country in a sweet spot, where it can benefit from Industry 4.0 development opportunities. Even so, its government has been working hard to fill up the gaps in the country's resources —notably its education and training systems which might need further optimization to expedite its process of manufacturing modernisation.

Industry 4.0 technology infrastructure — Foreign and local vendors offer Singapore's manufacturers with a full range of IT connectivity equipment, software, and services. Other converging technologies such as simulation technologies, 3D printing, automated robotics, and data analytics are also available as commercial products and custom solutions.

Government digital transformation initiatives drive and will continue to promote IT services growth in the foreseeable future. Singapore's manufacturers can also engage third parties, who provide IT project, managed, and support services for companies in manufacturing and related verticals such as logistics.

Government support for Singapore becoming a Smart Nation is the most reliable long-term driver of ongoing demand for the inputs of Industry 4.0 development. The government's continued support is a key factor in the nation's digital manufacturing initiatives in the private sector.

Telecommunications prowess — Singapore has developed the status of a world leader in telecommunications by building a high-quality network and managing progressive telecom regulations. These and the following resources promote digital manufacturing development:

- The now-completed rollout and deployment of Singapore's next-generation broadband network.
- Singapore's Smart Nation initiative now supports smart logistics projects. These projects extend connectivity and enable interoperability standards throughout the nation's manufacturing supply chain.
- A third Google data centre in Singapore.
- Partnerships between SingTel (Singapore's major telecom) with universities and manufacturers. For example, the SingTel FutureNow Innovation Centre helps enterprises accelerate their digital transformation.
- Other local telecom partnerships that set up IIoT infrastructure. StarHub is setting up a partnership to develop and launch commercial IoT applications and services in Singapore and M1 launched the first commercial narrow-band IoT network in Southeast Asia.

13.1.1 Education & Skills Development

The nation's government, businesses, and educational institutions are starting to work together in various policy initiatives and public and private partnerships. These policy programmes and collaborations provide essential resources for manufacturers wanting to modernise their thinking as well as their production methods.

The SkillsFuture Program — SkillsFuture Group, a community outreach initiative, is a national programme of education, training, and skills development. Designed to encourage lifelong learning in all Singapore's workers, the program is run in collaboration with Oracle Singapore.

New Learning Courses — Nanyang Technological University established the Centre for Professional and Continuing Education (PaCE@NTU) in January 2019. A series of Industry 4.0-related seminars sponsored by PaCE@NTU complement the SkillsFuture effort to promote industry-relevant training programmes.

As part of the SkillsFuture seminar series, partners PaCE@NTU and Singapore's Employment and Employability Institute (e2i) run a series of Industry 4.0-related topics in eight domain areas. Participants complete three 2-hour seminars in the same topic area to earn a Certificate of Completion in Industry 4.0 skills development.

As Industry 4.0 technology trends drive changes in manufacturing and supply chain management, PaCE@NTU partners with Delta – NTU Corporate Lab. Their goal is to develop online training content that helps professionals and enterprises learn and collaborate in the changing manufacturing environment.

The partnership uses a mobile application learning platform, which creates real-life learning scenarios with the latest industry information and data generated from office and production use cases.

All Singaporeans and permanent residents are eligible for 70-to-90-per cent SkillsFuture Series funding. These subsidies apply to all mobile learning courses offered through the DeltaKnEW Academy mobile app. SMEs also qualify for up to 90 per cent subsidies of the course fee under the Enhanced Training Support for SMEs scheme.

13.1.2 Education in Advanced Manufacturing

Singapore's advanced manufacturing, continuing education, and training strategy includes initiatives designed to attract and retain talent and develop workers' skills. SkillsFuture Singapore developed the plan by consulting with government agencies, industry partners, labour unions, educators, and training providers.

Koh Poh Koon, Singapore's Senior Minister of State for Trade and Industry, has emphasised the importance of public-private partnerships to harness the potential of manufacturing-related technological advances. He noted that 17 firms, including auto manufacturer Rolls-Royce and local firms, are collaborating on the IoT Innovation platform developed by Singapore's Agency for Science, Technology, and Research.

The platform is part of an industry-driven consortium of end-users, solution providers, and researchers. Consortium members use existing technologies to create end-to-end IIoT solutions. They also develop new technologies with A*STAR Research Institutes and Institutes of Higher Learning in Singapore. SMEs and start-ups also benefit from this collaborative system when they upgrade their capabilities and position themselves as solution providers to consortium end users.

Industrie 4.0 Specialist Diploma — Singapore's first Industrie 4.0 specialist diploma is supported by a partnership between the Technical University of Munich in Asia (TUM Asia) and provider of equipment and solutions used in technical education.

The partners launched Singapore's first Industrie 4.0 Specialist Diploma in Advanced Digital Manufacturing in May 2018. German manufacturing experts, the key drivers of the Industrie 4.0 program in Germany, guide the program in Singapore. The curriculum equips participants with the Industry 4.0 knowledge, skills and hands-on equipment experience that they need to get and keep manufacturing jobs of the future.

Industry 4.0-related online learning — Online, micro-learning courses provided by Temasek Polytechnic teach participants core Industry 4.0 concepts and how to apply them at work. Tests, quizzes, and group discussions measure participants' ability to absorb and apply these basic concepts and communicate what they've learned.

Robotics research lab — The Harbin Institute of Technology Robot Group, one of China's leading robot manufacturers, joined Temasek Polytechnic to launch the TP-HRG Robotics Innovation Centre in Singapore. Opened in February 2019, the laboratory can be used by SMEs to test robotic solutions in a risk-free environment.

Onsite training services — Local engineering firm, Univac Precision Engineering, teamed up with Singapore Polytechnic (SP) to develop a customised set of onsite training modules. The engineering company plans to use the Industry 4.0 training materials to improve the skills of its workers and supervisors. There is a new wrinkle, however. Rather than have trainees head to the SP campus to attend courses, the training program focuses on workplace learning. Trainees spend just 15 minutes each week for three weeks to complete a module. The program offers 10 modules to help Univac employees prepare themselves for Industry 4.0. Only 7.4 hours of training are required for the entire introductory program. Company supervisors are also trained in coaching methods and workplace learning methods.

The "bite-sized training" approach has been validated in the United States and elsewhere as a useful training method. This alternative to traditional, sit-down training is ideal for manufacturers, who would otherwise have to deal with disruptive downtime whenever they send workers off to training.

New Industry 4.0 research laboratory — Computer and printer manufacturer, Hewlett-Packard, is collaborating with Singapore's Nanyang Technological University to set up a new research laboratory. Lab operations focus on developing advanced digital manufacturing technologies. The first of its kind for HP in Asia, the S$84 million lab's research will emphasise 3D printing, artificial intelligence, machine learning, cybersecurity, and product customisation.

HP and NTU will also develop educational curricula that support additive manufacturing product design. Topics will include data management and security, business models, and designing products for a positive user experience.

China-Singapore skills upgrade program — Singapore Management University (SMU) and Alibaba Cloud announced a partnership to provide more than 1,000 working professionals with a chance to improve their work skills. The series of courses, Alibaba's first outside of China, is offered through the SMU Academy, the professional training group of SMU.

Singapore Institute of Manufacturing Technology — The Singapore Institute of Manufacturing Technology (SIMTech) is a Singaporean research institute, part of Singapore's Agency for Science, Technology, and Research (A*STAR). SIMTech helps Singapore's manufacturing industry become more competitive by working with local companies to develop high-value technology and human resources. The institute works on projects with all sizes of companies in sectors including precision engineering, electronics, semiconductors, medical technology, aerospace, automotive, and logistics.

Training includes master-classes, graduate diplomas, modular programmes, and research into manufacturing processes and systems, planning and operations management. Graduate degrees supported by partnerships with foreign universities are also on offer.

13.1.3 Private & Public-Sector Funding & Support

Singapore bases its economic development on a proactive—some would say aggressive—strategy to attract foreign investment and trade. Singapore's open trade policies, favourable lending to foreign investors, simple regulatory system, tax incentives, and political stability make the nation an attractive investment destination. Here are the major programmes and funding opportunities for Singapore's Industry 4.0 development.

Singapore's Smart Industry Readiness Index — Launched in Nov 2017 by Singapore's Economic Development Board, this programme prepares manufacturers to adopt Industry 4.0 technology and practices.

The index is an Industry 4.0 development tool designed to guide Singaporean manufacturers through various Industry Transformation Maps. The index was developed by a partnership between the EDB and global testing, inspection, certification, and training provider TÜV SÜD. SSIRI content and methods were validated by an advisory panel consisting of academics and industry experts. The index and transformation maps were designed for manufacturing companies of any size and level of business maturity.

Prioritisation matrix — The EDB and its partners also developed a useful tool for companies unsure of how to move from data gathering (SIRI) to developing step-by-step Industry 4.0 strategies. The tool was launched by Senior Minister of State for Trade and Industry, Koh Poh Koon, at the Hannover Messe tech show in April 2019. The new Industry 4.0 Prioritisation Matrix helps manufacturers make the important leap from asking, "What do we need to do to start our Industry 4.0 efforts?" to "What company-specific series of steps must we take to make progress?"

Productivity Solutions Grant — Coordinated by the Ministry of Trade and Industry's grants management office, this incentive programme helps companies automate existing processes and improve productivity by using advanced technology solutions. The PSG subsidises up to 70 per cent of project costs of industry-specific manufacturing, engineering, and logistics solutions. Cross-sector solutions such as data analytics, inventory tracking, and financial management also qualify for programme grants.

The grant programme is available to all SMEs that are registered and operate in Singapore. At least 30 per cent local shareholding is required, and solutions developed must be used in Singapore.

Partnership for Capability Transformation scheme — Administered by Enterprise Singapore and the EDB, the PACT scheme fosters business-to-business collaboration between Singapore's SMEs and larger enterprises. The PACT concept involves a lead enterprise forming mutually beneficial alliances with smaller firms.

The program supports up to 70 per cent of qualifying costs of partnerships participating in capability and business development projects. Capability development plans include technology improvements in supply chain, cooperative innovation, and knowledge transfer projects. Business development activities include resource sharing or building alliances and consortia.

Foreign direct investment — In Singapore, encouraging foreign direct investment is a pillar of government support. The 2018 UNCTAD World Investment Report indicates that Singapore was the world's fifth-largest recipient of FDI inflows in 2017. However, 2017 inflows fell by roughly 20 per cent compared to the previous year. (This reflected global downward trends in trade generally.) The United States, British Virgin Islands, Cayman Islands and the Netherlands are Singapore's most reliable investors.

13.2 Malaysia

In a 2018 survey conducted by Forbes Insight and Deloitte, Malaysia ranks high on the list of Industry 4.0 readiness. Even with Malaysia's relatively advantaged position, current trends indicate that it's not going to be smooth sailing for the country's early adopters.

The reasons include a basic lack of knowledge and understanding of digital manufacturing and several deep-seated business attitudes. In the awareness survey, about 30 per cent of executive respondents admitted that their company lacks a detailed understanding of Industry 4.0 technologies, a business case, and leadership vision. These factors and what respondents described as an emphasis on short-term returns will make longer-term, Industry 4.0-based revenue growth less likely.

Malaysian manufacturing executives consider Industry 4.0 technology and ideas to be excellent tools—for defence. Few companies, however, seem eager to use converging technologies and new practices to engage national and cross-border competitors. Finally, respondents referred to a bad case of analysis paralysis—too many Industry 4.0 technology choices and not enough procedural guidance to benefit from them.

These potential roadblocks to Industry 4.0 development pose serious challenges to Malaysian manufacturers. However, there are plenty of resources available for their immediate use.

13.2.1 Availability of Convergent Technologies

Technology availability and cost are key drivers of Industry 4.0 adoption. As Malaysia's relatively high standing of Industry 4.0 readiness indicates, manufacturers can acquire a wide variety of advanced technologies and related services as in-country offerings. These include:

- Robotics
- Cloud-based data storage and analysis services
- IoT internet-connected devices
- Simulation (AR and VR) software and equipment
- Additive manufacturing (3D printing) equipment and software

However, to some Malaysian manufacturers, especially SMEs, there's a potential problem. The origin of these services can determine whether business owners can afford converging technologies. Malaysian manufacturers have had to depend on foreign-based, relatively expensive technical and project management companies to upgrade their production systems. Some owners engaged overseas companies to buy equipment, software, and consulting services. Others signed up with foreign companies that operate offices in Malaysia.

In interviews, Malaysian-owned factory owners mention that they would prefer local, less expensive technology alternatives. Recent improvements in the availability of Industry 4.0 technology products and services are giving Malaysian manufacturers their wish.

ICT infrastructure — Although Malaysia is strongly committed to developing a technological society, the country's fixed broadband market remains underdeveloped. The country's fibre-based, high-speed broadband networks have been improving, but progress has been slow. Flat growth in the fixed broadband subscriber market has been the story during the past five years. The causes include minimal increase in the number of fixed lines and growing demand for mobile broadband services. Analysts expect this slow growth to continue until at least 2023. The recent announcement to roll out 5G network by major telecommunication operators in October 2019 might mitigate the communications and connectivity issues that faced by industries.

13.2.2 Education, Training, & Skills Development

"Changing" is the word that describes all levels of Malaysian educational institutions. There's a new Education Ministry created from a two-department merger. There's much more autonomy for public universities, a persistent debate about whether to run a single- or multi-stream education system, and talk about whether Malaysia can afford free education at any level. A brief profile of Malaysia's schools and universities includes:

- Primary and secondary school systems of 10,000 schools, half a million teachers, and five million students.

- Post-secondary education system of 20 public universities, more than 500 private institutions, 36 polytechnics, and 100 community colleges. Made up of more than 70,000 academicians and 1.2 million students (including 170,000-plus international students) attend these schools.

University education — In Malaysia, the ministerial rhetoric about higher education is changing. Once regarded as a sanctuary for knowledge transfer, the government is urging universities to take on a more modern agenda. Now, universities are told they must become suppliers of workforce talent and "one-stop knowledge centres for firms, industry associations, government agencies, and community organisations." The position of universities in society is at the forefront of solving problems for industry and the community.

Pioneering students — At least one Malaysian university, Asia Pacific University of Technology and Information (APU), has answered this ambitious call to action. No longer confined to gaining knowledge in classrooms, the school's pioneering students enter the Industry 4.0 era by experiencing blended learning. APU is a place where business professionals working in smart factory offices of the future can polish their knowledge.

The university develops Industry 4.0-relevant programmes by fusing practical, hands-on experience with academic and theoretical knowledge. For example, finance students also learn Industry 4.0-related information such as Fintech and cybersecurity technology and methods. APU's Industry 4.0 topics also include data analytics, business use of IIoT data, automated systems, and cloud computing.

University course designers work carefully with industry advisory panels to equip students with the latest technical and soft skills that future job markets will probably require. APU's industrial partnerships with accelerators such as GrowthX Academy and Supercharger also enrich standard classroom information with real-time data and scenarios. The university's Cybersecurity Talent Zone features military-grade, real-time cybersecurity monitoring software at the Cyber Threats Simulation and Response Centre, for example.

Ideally, the students' education will help them become competent technology professionals by the time they graduate.

13.2.3 Government Accelerator Programs

In its 2019 budget, the Malaysian government committed substantial money—RM5 billion—and talent to meet its Industry 4.0 goals. These programs have already started or are scheduled to begin in 2019.

The Malaysian Productivity Corporation — This organisation has already begun its Industry 4.0 readiness assessment programme. Its goal: to help about 500 Malaysian SMEs begin their digital modernisation journey by clearly defining their technology and business readiness.

Knowledge Resource for Science and Technology Excellence — This fund was established to boost the sharing of R&D resources between actors in the private sector and academe. Funded by the Ministry of Finance and managed by Bank Pembangunan Malaysia, the RM3 billion Industry Digitalisation Transformation Fund encourages industries to adopt Industry 4.0 technologies. The scheme provides companies planning to upgrade their production technology with a two per cent per year subsidised loan.

Centre of Excellence for Technology and Talents — As part of the nation's Industry 4.0 transformation, the centre organises education and training programmes that enable SMEs to fill the technology and knowledge gaps they need to benefit from new Industry 4.0 capabilities

Other inducements — To further sweeten the deal and speed up adoption, companies trying to embrace Industry 4.0 could also enjoy double tax deductions. This inducement applies if companies:

- Offer scholarships to students studying engineering and technology at the technical and vocational, diploma, or degree levels.
- Participate in the Dual-National Training Scheme for Industry 4.0 recognised by the Human Resources Ministry, or MIDA.
- Conduct relevant training programmes approved by the Human Resources Ministry.

#mydigitalmaker initiative — The Malaysian Digital Economy Corporation (MDEC), the Malaysian Education Ministry, and industry and academic partners collaborate to create the #mydigitalmaker initiative. This partnership helps participants to develop future digital creators by exposing young people to the creative and innovative aspects of digital technology.

The initiative's down-to-earth goals include equipping Malaysian youth with digital knowledge and skills. The program started adding computational thinking and computer science into primary and secondary schools. At the primary school level, digital information is added to existing subjects, such as mathematics and science. The Secondary School Standard Curriculum includes basic computer science subjects.

MDEC, which also provides computing experts to train university-level educators, recognises these institutions as teacher training centres. The agency selected 12 universities throughout the country to train teachers to integrate computing and computer science into their classroom for the new curriculum.

13.2.4 For-profit Training & Education Organisations

Educations are appearing more and more frequently online and in cities and towns throughout Malaysia.

Based in Petaling Jaya near Kuala Lumpur, the Industry 4.0 Academy focuses on helping recent graduates and experienced professionals keep up with the latest Industry 4.0 information. They provide intensive, 2- or 3-month technical courses online and in the classroom. The academy also offers one- or two-day seminars exclusively for managers and executives. Topics include the benefits of embracing Industry 4.0 technologies and helping businesses stay resilient in a fast-changing world.

The Kuala Lumpur staff of ExcelR Solutions Malaysia offers a 5-day, Industry 4.0 training programme. Its goal is to help manufacturers embrace Industry 4.0 principles and use them to find new paths to business growth. The sessions walk attendees through the different aspects of Industry 4.0 and show the problems, solutions, and benefits of real-world use cases.

13.2.5 Public-Private Partnerships

The Malaysian government's Economic Outlook 2019 report put a new emphasis on public-private partnerships that promote Malaysian entrepreneurship and investment. The government highlights these collaborations as part of a wider effort, which includes:

- Improving the nation's business environment by eliminating overly bureaucratic policies.
- Attracting high-value investments to improve economic development.
- Strengthening the manufacturing sector's technology, education, and data infrastructures.

The 2019 national budget allocates RM 210 million (more than USD 51 million) to support the transition to Industry 4.0 during the 2019-2021 periods. By October 2018, the Ministry of Finance had approved 45 of the 67 public-private collaboration projects it was reviewing.

Here are some examples of recent partnerships between Malaysian government agencies, manufacturing businesses, and educational institutions.

231

MIDA-Muehlbauer vocational training collaboration — The Malaysian Investment Development Authority (MIDA) and Muehlbauer Technologies Sdn Bhd are working together to enhance vocational training. The program also involves cooperation between Malaysian manufacturers and academic institutions.

Under the partnership, the German-based machine manufacturer contributed integrated automation machines worth RM200,000 to three universities (Malaysia Pahang, Teknikal Malaysia Melaka, and Tun Hussein Onn Malaysia) and the German-Malaysian Institute (GMI). Each participating university received a copy of the system mock-up. The gift also included 40 hours of training and support time for each school.

Centre of 9 Pillars R&D hub — The Malaysian Technology Development Corporation (MTDC) launched its industry 4.0 hub, Centre of 9 Pillars in November 2018. The centre is a fully equipped incubator and R&D space, which support the development of proofs-of-concept and prototypes at a fraction of the cost of imported solutions.

The centre will support the R&D efforts of digital manufacturing adopters and their partners, who include researchers, academicians, industry experts, and other business professionals. The centre hopes to position MTDC as the platform of choice for SMEs, who seek local Industry 4.0 solutions for their existing operations.

Malaysia Digital Economy Corp partnerships — The Malaysia Digital Economy Corporation Sdn. Bhd. (MDEC) is the government-owned agency tasked to lead the transformation of Malaysia's digital economy. Started in 1996, MDEC plans to help Malaysia develop into a fully digital nation by 2020. MDEC engages Malaysian companies and educational institutions in partnerships that provide in-country resources for organisations moving toward modernised manufacturing. These entities include:

- Axiata partnership — A strategic collaboration between MDEC and Axiata Group Bhd (Axiata) establishes public-private collaboration meant to boost the nation's cybersecurity industry.
- #mydigitalmaker — This partnership provides short courses and certification programmes on programming, coding, digital manufacturing, and other digital topics through their Continuous Professional Development Centres.

Advanced robotics institute — Malaysia Automotive, Robotics, and IoT Institute (MARii) ease the transition from traditional manufacturing methods to those based on digital manufacturing skills, knowledge, and mind-set. MARii launched its Industry 4.0 programme in 2016 by focusing on the adoption of digital manufacturing in the automotive sector. The agency's Industry 4.0 training programme began in 2018. Since then, the programme works with multinational companies and SMEs that want to expand their business into the global value chain.

Under its old name of the Malaysia Automotive Institute, MARii developed and delivered technology adoption programmes for the automotive industry since 2014. The programmes have included:

- Design Engineering and Prototyping
- Manufacturing Execution Systems
- Computer-aided Engineering
- Automotive i-Cloud Computing
- MARii Automotive Garage Information System

The programmes enable the adoption of technologies used in manufacturing processes all along the value chain, from product design to after-sales activities.

13.2.6   Government Funding & FDI

The Malaysian government has made provision for a wide variety of Industry 4.0-related grants, loans, and investment programmes.

MITI investment programme for SMEs. In the 2019 budget, the national government allotted RM210 million (USD50 million) to support the migration of SMEs to Industry 4.0 technologies in the period from 2019 to 2021.

In 2Q2019, the Ministry of International Trade and Industry (MITI) began the Industry 4.0 funding process sponsored by the National Policy on Industry 4.0 (Industry4WRD) initiative. Aimed at Malaysia's SMEs, the program started with the Industry4WRD Readiness Assessment. In this step, MITI officials identify participating companies that need capital to adopt Industry 4.0 technology and practices.

Under the Industry4WRD programme, MITI will choose the approximately 500 companies to participate. By March 2019, the assessment programme had received 80 applications from Malaysian SMEs in various industries. Each eligible firm will have a recognised Industry 4.0 expert evaluate their readiness to adopt Industry 4.0. MITI officials will continue receiving applications throughout 2019.

Industry 4.0 grants and loan guarantees. In Budget 2019, the national government established these investment vehicles for Malaysian manufacturers engaged in Industry 4.0 improvements:

- RM2 billion allocated under the Business Loan Guarantee Scheme, which provides government guarantees of up to 70 per cent.
- An Industry Digitalisation Transformation Fund of RM3 billion, which includes a 2-per cent loan subsidy.

- A matching grant under the Domestic Investment Strategic Fund for investments that enhance smart manufacturing facilities.
- An extended incentive period for the Accelerated Capital Allowance for automation equipment purchased in the period of 2018 to 2020.
- An accelerated capital allowance incentive of up to 200 per cent for companies engaged in manufacturing and related services.
- A capital allowance incentive for ICT equipment purchased in the period of 2018 to 2020.

You can find more information about government-sponsored incentive programmes in "Moving Toward Industry 4.0."

13.3 Thailand

Regarded as an economic development success story, Thailand finds itself struggling to continue improving its productivity. Government officials look to Industry 4.0 development to sustain the country's economic successes so far.

13.3.1 Industry 4.0 Technological Resources

Thai manufacturers can find the robotics; AI, data analytics, and IIoT technologies they need to develop their smart factory and advanced logistics capabilities. However, the number and business maturity of Thai-owned companies that deliver these technologies vary by type.

Robotics hardware and software — Automation and robotics have played a growing role in the Thai Industry 4.0 era, especially in the country's automotive and electronics industries. Thailand's robotics imports are projected to grow by double digits over the next few years. The 2017 World Robotics Report by the International

Federation of Robotics also viewed Thailand as a growing market for industrial robots. Annual sales of robots are projected to increase by 89 per cent relative to their 2016 level.

Artificial intelligence software — Although Thai manufacturers have just begun to add AI technology to their operations, the country has already built a reputation as an enthusiastic AI user. Manufacturers can find plenty of international companies (Microsoft, Oracle, IBM, and Dell EMC, for example) and dozens of locally owned development houses and consultancies. Few if any of the Thai companies specialise in AI software development, however.

13.3.2 Internet Connectivity Infrastructure

During the past few years, the Thai fixed broadband market has witnessed reasonably strong growth from a relatively small base. However, fixed broadband growth in Thailand remains relatively slow compared to other developed Asian telecom markets. These constraints are the result of a limited number of fixed lines and mobile broadband growth. Over the next five years to 2023, growth for fixed broadband services is predicted to taper off, while major fibre-optics providers continue to roll out their networks.

13.3.3 Education & Skills Infrastructure

If you're looking for a phrase that summarises Industry 4.0-related developments in Thai education, it would be "a work in progress."

Persistent inefficiency in public primary schools, the need to modernise university curricula, and the need to light an Industry 4.0 fire under vocational training providers are common concerns of Thai business leaders and government officials. All parties concerned are aware of the problems. Many public policies and public-private partnerships have sprung up, and implementation of policies that might improve these problems has begun. However, the inbox of

education-related accomplishments is still empty. Concerned stakeholders are waiting for substantial progress to take place.
Industry 4.0-related progress is different in Thai universities, vocational schools, and for-profit training institutes.

Thailand's universities under scrutiny — Thailand's more than 300 universities will probably struggle with two major challenges. First, a rapidly ageing population and shrinking birth rates will deplete the country's workforce. These changes are anticipated to decrease the current figure of 50 million workers to 40 million by 2040. This adds pressure for educational institutions of all kinds to get workers ready for modern manufacturing jobs and careers.

Fewer students fill university seats — Of the 300,000 available seats in Thai university admissions systems, about 70,000 remain empty. The number of students at open universities fell by 50 per cent, while private university attendance plunged by 70 per cent. Naturally, school officials are wondering where future cohorts of university students will come from.

Suchatvee Suwansawat, CUPT chairman and rector of King Mongkut's Institute of Technology Ladkrabang, has a suggestion. Instead of attracting secondary school graduates, Thai universities should focus on retraining and re-skilling people who are already in the job market. This approach would fill empty seats and help working people survive the oncoming tech disruption.

Not nearly enough technical students — Thai educators face a formidable challenge, caused by unfilled demand for jobs that require education in technology, natural science, and engineering topics. Manufacturers will need more professionals and technicians familiar with AI and advanced (intelligent) robotics to fill Industry 4.0 jobs. Part of the problem with unmet demand for technology workers comes from Thai universities, which have been slow to acknowledge the lack of technology students.

Government officials are calling for education programs that support life-long learning, critical thinking and domain topics that help students get and keep more modern jobs. This broadly publicised policy points to a nation-wide problem. The government might make filling more technical positions "a call to arms." Thai students, however, have not been listening and traditional student career choices continue to make the job imbalance worse.

Righting the imbalance of social science students — About 70 per cent of Thai university students are social science majors, many of whom stay unemployed after graduation. The difficult part of university education reform is convincing some social science students to consider degrees in technical professions (engineering) and the natural sciences.

Science and Technology Minister, Suvit Maesincee, also recently called on Thai universities to keep pace with technological trends. He urged universities to overhaul their curriculum for each major to; keep subject knowledge up to date, close outdated majors, and open new ones that are consistent with the government's Industry 4.0 workforce goals. The minister proposes a 50-50 split of students pursuing social science and technical-engineering jobs.

Worries about competition abound — Thai universities are worried about competition from schools and training courses that use online learning technology. Educators are most concerned about IT companies such as Google and Microsoft, who now offer online courses that might become compelling alternatives to university-based technical education.

Alternative courses are indeed cheaper and easier to sign up for than conventional university degrees. However, their concerns might be misplaced. Unless private companies offer complete sets of courses— what one might call technical "majors"—it's difficult to see where the competition might lie. New training schools can teach IT literacy as well as Industry 4.0 subjects. However, training schools are unlikely to provide the deep knowledge in many scientific and engineering areas that is already part of a university's educational portfolio.

Vocational training schools get a boost — What about secondary school graduates or other students, who are not destined for university education? Private companies that offer Industry 4.0-centred mini-courses, seminars, and technical training programmes have appeared in cities throughout Thailand. Classes offering this new content join modernised vocational training classes in the education marketplace. They are all available, and thanks to government policies, many are more accessible to students than ever before.

Nationwide vocational training upgrade — Pearson, the digital learning company, has received an official government endorsement from the Thai Government. This testimonial permits Business and Technology Education Council (BTEC) qualifications to be delivered in every vocational and higher institution in Thailand, in public and private sectors. The pilot implementation, which started in December 2018, is expected to expand BTEC programs across 800 vocational institutions in Thailand.

13.3.4 Funding Programmes & FDI

Funding SME participation in Industry 4.0 programmes and nurturing future FDI both present potential difficulties for Thai manufacturers.

A bumpy road ahead for incentive schemes — In the past 40 years, the Thai government has offered various tax, customs, and employment incentives under different government investment schemes. Introduction of the Industry 4.0 framework gave the Thai government opportunities to continue this policy. Offered through different government agencies, these plans are designed to attract multinationals looking to set up smart factories in Southeast Asia.

There's a potential problem, however, the success of government incentives, in general, is still open to debate. Members of the international community have concluded that these initiatives could erode the tax bases of countries, which offer competing incentives in a classic race to the bottom.

The OECD report, Inclusive Framework on Base Erosion and Profit Shifting, establishes a case against aggressive incentive plans and provides an example of this downward spiral.

Approved under royal decree on 10 October 2018, the new International Business Centre (IBC) scheme addresses concerns set out in the plan's framework by providing less generous benefits than earlier plans. Under the IBC plan, entities that currently receive benefits under these initiatives will continue to do so under existing conditions—until their current status expires. When that occurs, multinationals must apply to become IBCs. (This is where the conditions and choices that a multinational must consider qualifying as an IBC get very complicated. An entity's best bet is to refer to these plan details.

FDI in Thailand, 2018 — Thailand is an emerging economy that's very dependent on exports, which account for more than two-thirds of the country's GDP. This makes reforming its FDI infrastructure a high-stakes game, which must be taken seriously. The Thai government is making substantial improvements reducing red tape. Foreign-based businesses can now start their operations in as few as four days, down from an average of 28 days.

Changes in government policies have strengthened the rights of borrowers and creditors. Legislation requiring companies to appoint independent members of the board of directors hopes to clarify corporate governance and ownership controls. These and other policies have dramatically improved Thailand's Ease of Doing Business Index ratings in 2017 and 2018. A wobbly currency, however, still makes overall investment attractiveness a bit uncertain.

FDI role in Thailand — Foreign direct investment is an important element of Thailand's economic development, and the country is one of the major FDI destinations in ASEAN. After several consecutive years of decline, FDI inflows have largely recovered, increasing 3.7-fold from 2016 to 2017, reaching a total of USD7.6 billion. This recovery is due to increased investment by EU countries and strong inflows from other ASEAN countries and Japan.

In 2018, the total value of FDI in Thailand increased by 15 per cent, reaching USD219 billion (51% of the country's GDP). Japan and Singapore are by far the largest investors in Thailand and account for more than half of FDI inflows. Taiwan, the Netherlands, Germany, Switzerland, Mauritius, and the United Kingdom are also among the major investors.

Several important pieces of legislation established or expanded economic initiatives throughout the country. The Investment Promotion Act offered more incentives to invest in advanced technologies and research and development. The Eastern Economic Corridor (EEC) Act offers investors in the EEC benefits, such as tax subsidies, the right to land ownership, and issuing visas.

13.4 Vietnam

Like some of its ASEAN-6 peers, Vietnam isn't entirely ready to make the most of Industry 4.0 development. Availability of technology and other supporting elements are unevenly distributed throughout the country. Here's a profile of the resources that could support Vietnam's digital manufacturing effort, now and in the future.

13.4.1 Internet & Communications Technology

In Vietnam, the availability of Industry 4.0 support technologies is a mixed bag, stronger in robotics and softer in connectivity infrastructure. As in all matters technological in Vietnam, the country's internet and communications infrastructure is young but growing quickly. Although infrastructure growth is travelling in the right direction—up—several significant structural problems risk the delivery of connectivity resources to the entire country.

Internet connectivity — Not all parts of Vietnam are ready for Industry 4.0. There are notable differences in internet use rates in rural and urban areas. According to the government's plan, by the end of 2019, 3G/4G networks will connect 95 per cent or more of residential districts. Vietnamese MNOs predict average urban downlink speeds of 4 Mbps and more in cities and towns. Rural areas will have 2 Mbps speeds, but these values are dwarfed by the rates delivered in Singapore (60 Mbps) and Malaysia (30 Mbps).

Other structural problems threaten the delivery of reliable internet service. The market includes internet providers, which all use a single gateway. Any issue at the signal source could cause problems in signal delivery over vast areas. Internet connectivity in Vietnam generally relies mostly on the submarine Asia-America Gateway cable. However, MNOs are adding new internet capacity to the Vietnamese system with new cables. A USD450 million undersea cable that connects Vietnam to its neighbours started operations in early 2017.

Broadband access — Access to fixed broadband services has been built on top of an extensive digital subscriber line (DSL) network. Recent focus has been on fibre-based services. The penetration of fixed broadband services in Vietnam remains low, due mainly to a limited number of fixed lines and the dominance of mobile broadband services. Analysts expect rising growth in fixed-line services and continued strong growth in fibre broadband access through 2023.

Ho Chi Minh City is set to become the first smart city in Vietnam. There, development will focus on cloud computing infrastructure, big data services, new data storage facilities, and security monitoring centres.

Narrow-band IoT — Narrow-band Internet of Things (NB-IoT) technology exploits Long-Term Evolution for Machine (LTE-M) technology, which enables battery-run IoT devices to connect directly with a 4G network without a connection portal. By using NB-IoT, users can take full advantage of existing 4G LTE infrastructure.

242

Although NB-IoT has been launched in Vietnam, until recently, Viettel, the national MNO, had been unable to attract much interest from businesses. This response changed markedly when Viettel announced its successful implementation of the technology at the recent Mobile World Congress 2019. Now, manufacturers seek many practical NB-IoT applications.

In early December 2018, Viettel activated 30 transceiver stations in Hanoi, which offered NB-IoT services. This was the first platform in Vietnam to deliver a commercial IoT network. The whole platform and related facilities will be available to the general public in the first half of 2019.

13.4.2 IT Products & Services

One could say that the Vietnamese IT sector was slow to start but it is now growing quickly. According to the Vietnam Software and IT Services Association, the Vietnamese IT sector has grown steadily in the past few years. Most activity in the Vietnamese IT sector is based on foreign investment. Major multinational companies such as Samsung, Nokia, and LG continue to expand their operations in Vietnam, thanks to reasonable labour costs and favourable tax policies. Major global corporations also pour billions of US dollars into building the beginnings of a thriving IT service industry.

Meanwhile, Vietnam's IT industries have been moving up the value chain, offering more value-added jobs and services.

A growing IT services market — Demand for software as a service (SaaS) and other cloud-based offerings are expected to increase as Vietnam's internet infrastructure continues to grow. Outsourcing firms in Vietnam include Digi-Texx, Harvey Nash, SPI Global, and Swiss Post Solution/GHP Far East. These and other companies compete in a vigorous market.

A limited pool of IT services — Locally-based IT services can be hard to find. Lower-value programming and testing services are available, if someone wants to build something from scratch, they must be mindful that the in-country supply of advanced design and programming talent is still limited.

The limited pool of skilled IT techs and deployment specialists makes it difficult for companies to take on large projects. Few vendors in the Vietnamese IT sector have more than 1,000 workers. Many have fewer than 500. So, software development, testing, maintenance, customisation, and support services are readily available. Few Vietnamese vendors, however, can provide their clients with end-to-end services or specialised skills.

Not enough university-trained technical graduates — Analysts and foreign investors call on the government for a continued focus on workforce training. Established Vietnamese universities cannot produce enough highly skilled graduates to meet likely future demand.

Currently, universities offer only one-to-three-month internships in the technology topics useful to Industry 4.0 development. These are not long enough to train prospective IT staff members in the skills they need. Industry players in Vietnam say that a focus on vocational training will help with the shortfall. However, this alternative doesn't address the need for more degree-holding engineers, scientists, and other technical specialists.

A bright note in the labour market — There has been an increase in the IT workforce due to the opening of new universities and colleges. These trainees and graduates add to the cadre of high-tech manufacturing specialists entering the country from overseas. They've moved to Vietnam to work for the local operations of multinational companies such as Intel, Samsung, and LG.

Limited access to advanced software resources — Data analytics and visualisation applications, as well as Blockchain and simulation software, are expected to have critical roles in running smart factories of the future. The question is, can manufacturers operating in Vietnam get access to these resources today?

Possibly, but only from Microsoft, Accenture, and other foreign companies operating in Vietnam. The Vietnamese software sector is still relatively modest. According to the EU-Vietnam Business Network, locally-owned software vendors account for 75 per cent of the market, which is dominated by low-cost products. Despite its small size, however, the sector has seen growth in recent years. Local software producers mostly provide their products to government agencies and SMEs. Larger Vietnamese companies generally look to multinationals for their software.

Analysts predict that Industry 4.0 modernisation will occur as a result of breakthroughs in AI, robotics, the Industrial Internet of Things (IIoT), and 3D printing technologies. An important part of Industry 4.0 readiness will occur when these technologies become easily available and affordable to Vietnamese manufacturers. The question is, has that day arrived?

Robotics — In its Industry 4.0 ambitions, Vietnam plans to make the considerable leap from traditional automated production systems to connected systems that process and continually guide the flow of data. Vietnam wants to industrialise and modernise its economy by 2020. To that end, the government urges businesses and educational institutions to actively pursue science and technology innovation and attract foreign investment in digital manufacturing technologies, especially robotics.

Local and foreign manufacturers doing business in Vietnam have easy access to robotics technology. All the familiar giants of the robotics industry —ABB for example— operate in the country, but what about local participation? Although Vietnam still doesn't have its own robotics industry, several local enterprises are creating useful products for the market.

- The Robot3T Group designs, develops, and manufactures high-precision robotics, motion controls, and integrated, industrial automation solutions. The group's hardware and software products range from tool automation modules and associated controls to complete integration solutions.
- VN Robotics started in 2012 started with early financing from AKB Machinery, Logicom Vietnam, and other companies and individuals who participated in the Vietnamese electronics and automation industries. The company focuses on the field of automation. Its key products include industrial robot arms, model robots, teaching aids, self-propelled robots, and other automation devices.
- Airobot is a Hanoi-based company provides manufacturers with robotic system design, production, and integration services. Services also include repair and deployment of existing robotic systems and debugging and creating robotic control software.

AI software, a vital part of automated production processes, has had a different path to development.

AI applications — Vietnam wants to emerge as an AI leader in Southeast Asia and become a pioneer in AI products generally. According to the Boston Global Forum, this goal will be difficult to achieve. With low levels of AI startup investment, a weak AI technology community, and few skilled programmers, locally-developed AI products are unlikely for the foreseeable future.

The outlook for IIoT resource development in Vietnam is also grim.

IIoT resources — IIoT devices in the context of smart manufacturing enable factory managers to make operations more efficient and make better-informed decisions.

IIoT solutions are available in Vietnam. However, the vendors who offer them tend to be well-known foreign operators such as Microsoft, Oracle, ABB, Avnet, and FPT Software. To date, Ho Chi Minh City-based TMA Solutions is the only Vietnamese-owned IoT outsourcing.

13.4.3 Education & Skills Development

Education in Vietnam is a tale of two systems: progress and successes that look good on paper and the real-life difficulties that create a very different reality. With a population of approximately 95 million, Vietnam is becoming an important market for global technology vendors and home-grown entrepreneurs. Despite Vietnam's achievements, however, finding education and training that support digital manufacturing is still problematic.

In 2016, the Vietnam Software Association (VINASA) reported that Vietnam would require approximately 400,000 IT workers for the 2016-20 periods. However, the country's 290 universities and junior colleges and another 150 training establishments can supply only 250,000 skilled workers during that period. Le Xuan Hai, chairman of Vietsoftware International, recommends that companies should engage in in-country training programs that enable workers to contribute to the market.

13.4.4 University Education

Figures for 2017 show that approximately 200,000 Vietnamese graduates with a bachelor's degree remained unemployed. Colleges and universities, especially those with an emphasis on STEM subjects, are opening throughout the country.

The Vietnamese government has been pursuing educational reforms for more than a decade. Nevertheless, the limited availability of modern educational opportunities is due to:

- Faculty members, who are underqualified and unable to meet student needs.
- Universities that stubbornly remain disconnected from the country's job markets.
- Topics and teaching approaches that emphasise theory but don't reflect the knowledge and skills that businesses want their employees to have.
- Recently graduated teachers who are trained to use outdated teaching methods.

This reality runs contrary to the bright picture that the government encourages. The result: despite the effort and funding, there are not enough programs to educate technical and manufacturing professionals.

However, the government isn't letting the deplorable state of higher education fester. Instead, their response to problems in higher education is to strike out in a new direction.

A new approach to university education — The Fulbright University Vietnam combines American academic traditions of liberal education and local innovations that concentrate on solving national challenges in higher education. Launched in September 2018, faculty and students collaborate to design the school's curriculum before beginning to teach in their inaugural 2019-2020 academic year.

The university emerged from the Fulbright Economics Teaching Program, a partnership between the Harvard Kennedy School and the University of Economics Ho Chi Minh City. Classes are due to begin later in 2019.

There's no way to know whether the university's new approach will work or how long it will take for other universities to follow the Fulbright example. With even the most favourable outcomes, there's still no solution to providing the professionals who can fill positions in smart factory shop floors, offices, and design works.

13.4.5 Government Funding & FDI

Good things are happening to the Vietnamese economy. These include a young, growing workforce and a manufacturing infrastructure that seems willing to adopt new technologies. However, the inevitable question, "Where's the investment money going to come from?" arises, and manufacturers must find the cash. Funding opportunities are available via government incentive programs and FDI.

Vietnam's Industry 4.0 financial environment — Financial trends in Vietnamese manufacturing companies point to a potentially serious roadblock to Industry 4.0 development: the lack of self-funding. Results of a 2018 survey by the Vietnam Chamber of Commerce and Industry (VCCI) point out that:

- Only 6.6 per cent of domestically invested manufacturers surveyed stated that they have enough resources to shift from their current out-of-date production system to a cutting-edge one in the future.
- Approximately 35 per cent of surveyed enterprises said that a lack of resources will enforce a step-by-step approach to Industry 4.0 development.
- Approximately 28 per cent of respondents revealed they are mobilising capital and resources to finance impending Industry 4.0-releted changes.

Fortunately, government policies provide solutions to some Industry 4.0 funding problems.

Policies that promote IT development — Vietnam aims to develop a legal framework that will accommodate the new technologies of Industry 4.0 and accelerate public-private partnerships. According to the VINASA, the Vietnamese government has adopted policies creating favourable conditions for IT development. These policies have included:

249

- Politburo Resolution 36-NQ/TW, which involves financial incentives that will increase IT application and development.
- Prime Ministerial Decision 392/QD-TTg of 2015 approves a target program for development of the IT industry by 2020.
- Resolution 41/NQ-CP, a tax incentive provision to promote IT application and development to improve competitiveness and attract investment in the manufacturing sector. The resolution provides preferential terms on corporate and personal income taxes. For example, new IT projects at companies with more than 1,000 full-time employees enjoy an annual corporate income tax of 10 per cent for 15 years.

Vietnamese Government Decree 13 makes the development of science- and technology-based companies much more attractive than previously. The decree describes how businesses can obtain Scientific/Technology Enterprise Certification. This policy enables businesses to take advantage of Vietnam's strong government Industry 4.0 incentives. Qualifying businesses would enjoy incentives such as corporate tax cuts and exemptions, credit incentives, and reduced land and water surface lease fees.

However, no matter how potent Industry 4.0 incentives might become, trade policies are the most important government policies in Vietnamese economic development.

FDI trends — The Vietnamese economic model depends heavily on foreign investment and exports, especially to the United States and Europe. In recent years, Vietnam has been very committed to trade liberalisation. The country's greatest progress has been protecting minority shareholders and cross-border trade. The country is willing to create a more favourable environment for foreign investors.

Foreign direct investment into Vietnam rose by 6.2 per cent year-on-year to USD 4.12 billion in the first three months of 2019. Pledges of future FDI disbursements increased by 86.2 per cent from a year earlier to USD 10.8 billion. Discrete and process manufacturers are expected to receive the largest amount of this investment, about 78 per cent of total pledges.

Vietnam's Department of Foreign Investment calculated 2017 FDI at USD35.9 billion, a 42 per cent, year-to-year increase. Traditionally directed towards light industry, FDI inflows in Vietnam now turn towards heavy industry. Major investor countries included Japan, South Korea, and Singapore, with the manufacturing sectors attracting the most FDI. Disbursed FDI, which the Vietnamese government expects to continue to rise, now enables manufacturers to produce high-value-added products such as smartphones and tablet computers.

13.5 Indonesia

Indonesia's economy might be holding steady, but the country faces shrinking growth and more difficult foreign trade, which threaten to make the country less competitive. Now, the resources needed to run more efficient manufacturing and logistics processes are more important than ever.

13.5.1 Education & Vocational Training

As in other types of Indonesian infrastructure, the nation's educational systems are playing catch up with its ASEAN peers to become ready for Industry 4.0.

Despite a broad range of educational reforms started since 2005, Indonesia still struggles to provide its citizens with an inclusive, high-quality education. Improved teacher training standards and sizable boosts to education spending as a share of the national budget (20 per cent in 2018 as required by national law) have improved education markedly in the past 15 years. However, public school spending has stagnated over the past decade and remains well below the 3.6-per cent of GDP level that The World Bank recommends for emerging economies.

Literacy levels continue to be a problem, especially compared to other Southeast Asian nations. A World Bank analysis showed that 55 per cent of Indonesians who complete school are functionally illiterate compared with only 14 per cent of comparable students in Vietnam.

Government efforts at education reform — Government ministers and administrators are paying attention to the need for educational policies that integrate smoothly with longer-term employment strategy.

Ir. Darwanto, Head of the Secretariat of the National Professional Certification Board, addressed the educational requirements of Indonesia's future workforce. The plan he described, however, used a somewhat unfocused approach, stating: "re-designing the curriculum with digital as well as human resources and the need to identify digital competencies and investment requirements for the nation's future workforce." There's been little recent progress to increase the availability and quality of Indonesian public education.

Post-secondary education — Beyond a ninth-grade secondary education, Indonesian students can take several paths to advanced learning. Research institutes, polytechnics, universities, academies, and schools of higher learning all provide students throughout the country with face-to-face and digital learning methods. In 2018, 26.3 per cent of Indonesia's 6 million-student cohort of post-secondary learners are working on degrees in engineering, mathematics, and computer sciences, a 22 per cent increase since 2014.

The government's efforts to create an Industry 4.0-ready workforce have not been enough to counteract the effects of serious structural problems. Low tertiary attainment levels hover at 9 per cent, the lowest in ASEAN, and university graduate unemployment rates remain stubbornly high. Although the research output of Indonesian universities continues to grow, it's still low compared with that of other emerging economies.

252

Higher education — Schools of higher learning are the most common form of Indonesian higher education institutions (HEI). There were 2,488 HEI throughout the archipelago in 2018. Most are smaller, specialised private schools. Typically, advanced schools offer undergraduate programs in an area of study and lead to a four-year diploma or bachelor's degree. Some HEIs however, also offer graduate degrees. Polytechnics, on the other hand, generally offer shorter, employment-centred programmes. In 2018, there were 257 polytechnics in Indonesia.

In a 2019 interview, Director General for Science and Technology Institutions and Higher Education, Patdono Suwignjo, acknowledged that there aren't enough polytechnic schools to supply the Industry 4.0-literate workforce that the nation requires. Currently, these schools comprise only 6 per cent of Indonesia's HEIs. This level is far lower than that of polytechnic and vocational schools in advanced industrial countries.

The quality of the nation's polytechnic education as reflected in schools' accreditation is also a problem. Minister Patdono remarked that. "Indonesia lacks high-quality polytechnics. In fact, only 3 of the country's 257 polytechnics have earned an 'A' accreditation."

Vocational schools — Indonesia's vocational schools are also under public and political scrutiny. Like the country's universities, the quality and quantity of training in vocational schools are deemed insufficient to support a growing population and economy.

The major problem is alignment: Indonesia's educators and trainers provide skills that are out of date and fail to match the demands of the country's employers. This shortfall of trained workers is especially painful in the manufacturing sector, where employment skills play a significant role in Industry 4.0 readiness. The root cause has been a failure to keep pace with technology and new labour demands, even at the country's best vocational schools.

Reforming vocational education becomes increasingly important as the Indonesian economy reaches higher growth rates. The importance of this topic in the public consciousness is reflected in newspaper articles and national political debates. Pledging to emulate Germany's skills training system, the Widodo government has allocated USD1.22 billion in 2019. More than double the past three years' spending, this funding will affect 320,000 students at 14,000 vocational schools across Indonesia.

Distance learning — In a 3,000 mile-long archipelago like Indonesia, distance education is an attractive way to provide educational services to people in remote areas. The lack of electricity and internet infrastructures and low computer literacy slow educational improvements in some areas of Indonesia. Nevertheless, e-learning in Indonesia grew by 25 per cent between 2010 and 2015 and still has excellent growth potential.

In post-secondary educational institutions, distance education was pioneered by the public Indonesia Open University (Universitas Terbuka). This school grew from a small, fringe university when it was founded in 1984 to the 500,000-student mega-university it is today.

Its mission was to provide underserved populations in isolated regions with an advanced education. The adoption of distance learning in Indonesia is still modest. However, growing numbers of other HEIs have followed the Universitas Terbuka example. Universitas Terbuka delivers diploma, bachelors, and master's programs through distance and hybrid learning systems. The university uses media tools such as radio and TV broadcasts, videotapes, and increasingly, the internet.

13.5.2 Government Funding & FDI

By all accounts, 2018 was a good year for Indonesia. UNCTAD assigns Indonesia's FDI growth to economic policies enacted by the Indonesian government during the previous several years. The national government introduced stimulus packages that focus on (among other things) deregulation, interest rate tax cuts for exporters, and investment-related tax incentives in special geographic zones.

Even this bit of sunshine was blocked by financial clouds on the horizon. First, the government carried out a reform programme meant to liberalise the economy and reduce investment barriers. Shortly after, Indonesia fell to 73rd place of 190 in the 2019 Doing Business Index. Then, the nation's Constitutional Court granted regional governments more regulatory authority. This move could challenge ongoing improvements in the investment climate.

FDI trends — Foreign investment is the bulwark of Indonesia's economy. Positive FDI numbers go a long way to support shaky performance elsewhere in the economy. Unfortunately, the recent FDI news has not been accommodating. Based on data from the Investment Coordinating Board, FDI levels dropped 20 per cent year-on-year to USD 5.9 billion in 3Q2018. At a January 2019 conference in Singapore, there were some concerns about red tape and uncertain ROI for Indonesian projects scattered among the agreement with highlighted policies and predictions.

14 Guidelines for Industry 4.0 Practitioners

"Learning how to understand how technology evolves, using tools like a Technology Road Map, is what you need more than anything to ride on top of the tsunami instead of being crushed by it."

-Peter Diamandis

A readiness assessment can help you identify the starting point of your digital manufacturing journey. A digital manufacturing roadmap, the second step in Industry 4.0 development, is just the ticket to help you identify and prioritise all the actions along that journey.

Regardless of the project scope, adopting digital manufacturing technology puts a company through the wringer. So, why bother? Industry 4.0 is all about creating value. Technology researchers agree that converging operations and information technologies will create remarkable opportunities to create valuable new ways of making and selling goods and services.

14.1 Where's the Value?

The Industry 4.0 story boils down to three areas of potential value creation: vigorous revenue growth, stronger profits, and fewer (or less severe) risks. At a very general level, manufacturers engage in digital manufacturing for several reasons. They want to:

- Compete more effectively on a regional or global scale.

- Respond to growing customer expectations for speedy delivery, more convenient use, and more personalised products.
- Enjoy faster, more accurate decisions, which are enabled by access to timelier, relevant data.

So, digital manufacturing value can be found in faster time to market for more highly personalised goods and services. This is reflected in faster decisions and stronger business relationships built by customers, suppliers, and channel partners working as collaborators.

14.1.1 What's in a Good Roadmap?

Strategic roadmaps for Industry 4.0 should help today's manufacturers understand what digital manufacturing requires of them and the challenges that they might face. This approach involves carefully identifying and planning every step a manufacturing company needs to take—and the timeline, costs, and benefits associated with each step.

However, not all roadmaps are created equal. There's more than one "proper" type of Industry 4.0 roadmap and more than one proper way to create it. This chapter looks at strategic roadmaps, their contents, and what they can accomplish.

14.2 Building a Roadmap for Business

Creating a detailed strategic roadmap is the second step of digital manufacturing development. Results of a completed readiness assessment expose the gaps in capabilities, equipment, processes, and best practices that can promote optimum operations efficiency. Government agencies, trade organisations, and industry standards groups publish Industry 4.0 roadmaps. We'll start with the ones created closest to home, by individual manufacturing companies.

There's a knack to writing company-level Industry 4.0 roadmaps. It requires attention to a business' goals and vision at the planning stage and highly detailed documentation during the writing phase. Here are several thoughts and approaches gleaned from engineers, consultants, and government officials engaged in Industry 4.0 development.

Data and connectivity make factories smart — Making the most of digital manufacturing opportunities requires a goals-first approach. The high-level goal of an Industry 4.0 roadmap is to connect data flows that enable better decisions and make manufacturing processes more efficient.

The fastest way to achieve smart factories and Industry 4.0 business opportunities is through thoughtful data management — collecting, analysing, and documenting data connections throughout the manufacturing value chain.

You can make legacy equipment smart — Although pervasive hype suggests otherwise, even older equipment can get smarter. What matters is making sure that factory technology, no matter what its vintage, can connect with new and legacy equipment. Managers can always get value by integrating modern technology with selected old machinery. More importantly, machines configured with a mix of old and new technologies can help you make intelligent, data-driven decisions about factory processes.

It's the new capabilities that lie at the heart of what is meant by smart. A factory doesn't require the latest or most powerful technology to modernise its processes. Instead, any level of digital connectedness can help manufacturers understand how manufacturing processes work. It's the understanding, not the technology that enables the best decisions.

You can make legacy equipment more capable, too — There are many ways to develop more efficient processes without installing lots

259

of new and expensive equipment. Digital manufacturing process designers suggest that companies slow down at early stages of Industry 4.0 development and make the most of their legacy infrastructure.

Many businesses, especially SMEs, have older manufacturing equipment that doesn't integrate smoothly with IoT devices, 3D printing platforms, and other new technologies. For these companies, real-world Industry 4.0 involves making old technology work alongside the new. Sometimes, the only step manufacturers can take with old machinery is to connect it indirectly to monitoring software with a simple sensor. Though simple, this lower-tech solution still provides useful information.

Incremental change on a budget is possible — Screamer headlines describing Audi's new $1.3 billion, everything-connected factory suggest that a company must make wholesale improvements to derive value from digital manufacturing operations. However, it's not necessary to invest lots of development money all at once; one can achieve successful improvements in small stages, one process, assembly line, or product group at a time.

Before starting Industry 4.0 implementation, understand how to streamline existing data flows and improve current manufacturing processes in different parts or levels of factory operations. Then, improve as much of the data flow or process as you have time and resources for. This incremental approach makes it easier to document, test, and pay for new technology investments that will further digitise your operations.

As you can see, we're bullish on incremental improvements. This step-by-step mind-set colours our view of the how-to elements of roadmap building in general.

14.2.1   First-Things-First & Step-By-Step

An Industry 4.0 roadmap is a prioritised action plan that's customised for each company's business and industry. Consider and perform these essential planning, implementation, and testing steps during the roadmap writing process.

Planning — To start, develop a company-wide description of your Industry 4.0 future. Yes, this step is what a former American president once called "the vision thing." The only reason anyone creates an Industry 4.0 roadmap is to guide their company's implementation in a clear and structured way. Who knows what to do, until you know where you want your Industry 4.0 development to take you?

It pays to go slowly and develop a detailed description of all the main things you want to adopt and achieve—technologies, capabilities, processes, practices, and attitudes.

Again, there's no need to cover the entire company at once. Just carefully define the scope in terms of time and the RAMI 4.0 model. Then, just stick with the scope throughout the analysis.

Define your starting point carefully — It's difficult to overstate the importance of setting up a clearly defined baseline. We define "care" as a detailed status report of technologies, capabilities, processes, and practices. If you think that these items resemble the vision points described earlier, you're correct. Each item in the baseline description is a current-day match to each vision point.

This is the point where your readiness assessment results prove their value. A useful Day-One status description depends heavily on a realistic interpretation of your company's and industry's Industry 4.0 readiness.

Clearly define implementation goals — You've established qualitative results in terms of desired outcomes —your vision—and know where you stand in terms of readiness. These outcomes aren't goals, however, until you assign them target metrics and deadline dates.

Start by identifying the highest-priority goals as defined by your company's critical success factors and business goals.

Next, use the specific points of the RAMI 4.0 model to identify the scope of your improvements. Each scope item will help you describe the capabilities, tasks, and resources required to improve your project's process, assembly line, or product group.

Review and integrate business processes — To adopt digital manufacturing in your business, begin by reviewing the business processes you want to improve. Then, integrate the enterprise systems that support them. There's a rather good chance that your business models (and therefore, your current business processes) will change. It's essential that you review these processes with a clear sense of future business objectives. This won't be an easy task because implementing Industry 4.0 will radically change the flow of information throughout your company.

Look beyond the factory door — Industry 4.0 concepts will make you expand your view of the people and processes involved in digital manufacturing. That's why, when planning your Industry 4.0 roadmap, it pays to view key suppliers as an extension of your organisation. For example, what do you do if you want to speed up time to delivery and your current suppliers are slow to deliver their goods? You might want to redesign your supply process or get a new supplier.

Don't forget the human element — When talk turns to Industry 4.0, people usually focus on the technology. Design and process improvements might involve learning how to use new hardware and software, but tech isn't the whole story.

Specifying digital manufacturing improvements also requires knowing the right people. These are IT and OT specialists, who know the relevant technologies and have experience deploying them,

262

successfully and securely. Whether you choose your in-house IT department, vendors, or an outsourced IT provider — service providers should have significant experience in orchestrating data flows across operations and businesses. It's a good idea to engage and include them in your improvement efforts at an early stage.

Collaboration and communications matter — Collaboration between people in previously isolated functional areas is a hallmark of Industry 4.0. Building a functioning system that blends software and information with physical processes requires the alignment of information and operations tech. When planning your roadmap, make sure that these departments will be in constant communication and committed to learning from each other. Without solid C&C, the data will not flow.

Beware of data overload — Your smart factory and offices will probably create a mountain of data and require high-speed, high-volume analytics. This will likely add new equipment, computing, and process requirements to your operations. You could connect virtually everything and everyone to the internet. However, it pays to be cautious about what you connect. Keeping data sets manageable keeps time-to-decision short and costs under control.

Understand and address security risks — IT security environments are experiencing a blur of cybercrime, hacktivists, and intellectual property theft. There's a good chance that digital manufacturing projects will expand your operation's attack surface. This reality emphasises the need to evaluate potential security and operations risks. As in all Industry 4.0 planning matters, start with the highest-value or highest-priority scenarios. Then, get an IT security assessment to help you identify risks and propose solutions.

Completing the planning stage identifies what you want to improve, when you need it done, and how you define success.

14.2.2 Building your Roadmap

The next step in roadmap construction involves filling the goal framework with the capability, task, and resource details required to carry out each improvement to the desired level of maturity. You might recall from our discussion of maturity and readiness in Chapter 8, that an understanding of process and tech maturity plays a vital role in readiness assessment. Maturity information also plays a crucial role in specifying improvements in a smart manufacturing factory and business.

Filling in the details — First, gather relevant topic categories from the Industry 4.0 framework of your choice. Checking the framework's list of technologies and smart factory capabilities will help you avoid forgetting anything pertinent to your project.

Next, gather RAMI 4.0-based scope information described in the planning stage and maturity-related criteria taken from the readiness assessment. This and your process design team's knowledge of your facility should provide the information you need to build a complete set of tasks.

Make continuous improvements — You'll never be done with adding, deleting, or changing performance criteria or their values to your roadmap. As you receive feedback or pilot results, expect to tweak your roadmap with continuous refinements. Remember, the more flexibility you build into your methods, the more easily you can make changes during the pilot stage.

14.2.3 Testing

When you've filled the equipment, process, practices, and attitude details into your roadmap framework, there's still testing work to be done.

Dream big but start small — Make this your mantra throughout the pilot process. The trick is to start with a limited but clearly defined part of your operations, quickly tests it, and quickly scale up whatever works. The main idea is to start and finish the testing phase before the pilot inertia sets in. This well-documented phenomenon is often responsible for implementations failing to progress beyond pilot tests.

Now that we've described how a roadmap can be built at the company level, we'll show that all roadmaps are not the same. They vary by purpose, scope, and content.

14.3 Industry 4.0 Roadmaps for Industries

So far, we've described roadmaps as planning and decision-making aids that describe and guide complex, phased processes at individual companies. These carefully tailored documents provide a structured way to enforce "forget nothing" and first things-first approaches to Industry 4.0 implementation.

Considering we already know the must-have information (how to gather, derive, and use data taken from different sources) for a first-rate roadmap, you may wonder what is in those documents?

Well, it all depends on what type of roadmap you're building. If we find one on the internet, Making Indonesia 4.0, for example, it's probably been created by a national or regional government agency. If so, its purposes are education and persuasion. This is quite different than the step-by-step procedural document described in the previous sections. So, comparing different Industry 4.0 roadmaps reveal the ideas, content, and intent of these important how-to documents.

14.3.1 Roadmaps for Discrete Manufacturing

Different Industry 4.0 roadmaps have different content and objectives. At the company level, practitioners identify, prioritise, and describe the specific processes and resources needed to make manufacturing and its related business operations more efficient throughout product life cycles and value chains.

265

Discrete, especially complex discrete, manufacturing sectors such as aircraft, ship building and medical devices require deep engineering efforts and operational excellence to handle multiple subassemblies throughout the long production cycle. It's a mandate for manufacturers to manage complex prototypes, sophisticated designs and long production process. Hence the classic operations divide the whole production into bits and pieces – which own separate eco-system – to support the end outcome. A typical example is automotive manufacturing which is categorised into four major stages: stamping, welding, painting and assembly. This way the focused quality control and mass production will be achieved. On the other hand, the common practice with the discrete manufacturing raises the challenge when moving towards industry 4.0 due to the high cost associated with shop floor connectivity and feedback control - which relates to the close loop optimization.

Hence manufacturing and mechanical engineers play a key role by identifying the key pain points and prioritizing them based on the KPIs targeted to achieve. These activities include understanding the status quo and specific goals, and further design the roadmaps to guide their colleagues, suppliers, and partners in the detailed steps of their collective digital manufacturing journey. As described in this and earlier chapters, company-level roadmaps encompass

- Identify key concepts and standards that guide implementation efforts.
- Collect process-level, company information from other sources—readiness assessments, conceptual models such as RAMI 4.0, and many relevant processes from their factories and offices—to provide the necessary details.
- Define the process-level scope and provide a framework structure with ideas taken from the RAMI 4.0 model.

- Add readiness-related maturity information to define the objectives of their process improvements as well as the resources and methods needed to implement them.

This approach to describing and implementing company-specific goals was designed for businesses in discrete manufacturing industries. Here are some use cases that have been planned and conducted:

- OEE optimization with the AI/Machine Learning and advanced analytical tools.
- Quality control along the production line based on IIOT data collection and condition monitoring.
- Interoperable data flow enabling fast response from corporate resource planning to assembly line scheduling, vice versa.

14.3.2 Ind. 4.0 Roadmaps for Process Manufacturers

In today's process manufacturing environments, there's growing awareness and practice of process control optimisation and value-added production. This understanding includes a focus on applying Industry 4.0 concepts to process automation. Forward-looking process manufacturers want to achieve end-to-end integration of automation, business information, and manufacturing execution to improve all aspects of production and commerce.

Merging the old with the new — Collecting production-related data for the process industry is not new. However, until recently, the strategy of merging OT and IT data and converting it into actionable information was seldom used in the process industry.

In the mid-2010s, interest in Industry 4.0 ideas for process manufacturing use cases replaced scepticism that had been common in the industry at that time. That was when process manufacturers realised that they were already using some Industry 4.0 and IoT concepts and technologies.

When it comes to implementing digital manufacturing, the process industry has a significant, built-in advantage over discrete manufacturers. For decades, process manufacturers have collected much of their data via DCS or PLC and SCADA systems. Also, the use of Manufacturing Execution System (MES) is common. That reduces up-front investment costs of IoT technology and their related risks.

Distinguished authors — Written in Germany in 2015, the Process Sensor 4.0 Roadmap describes an opportunity for manufacturers to optimise process control and value-added production. This implies a wider use of embedded, networked sensors that communicate throughout existing production operations.

Process Sensor 4.0 was written by NAMUR, an international association of automation technology users. They were joined by VDE, the Association for Electrical, Electronic, and Information Technologies, one of Europe's largest technical-scientific associations. Prominent leaders in the German process industry, (ABB, BASF, Bayer Technology Services, Siemens, Fraunhofer and others) rounded out the roadmap's authors.

Connectivity, integration, and interoperability — Process roadmaps resemble those used in discrete manufacturing businesses in several important ways. In both cases, the ultimate idea is to connect process applications with IoT-enabled devices. The goal: to ensure better communication between manufacturing and business systems, the supply chain, engineering functions, and planning processes.

The sensors are implemented by using embedded, internet-connected IT. Adding IoT technology expands the scope of production processes. Once rigid, hierarchical production systems focused entirely on a single production line expands into networked systems that communicate with each other and business processes within and beyond the manufacturing company.

The process roadmap focuses on achieving greater efficiencies with sensors, which have embedded intelligence, communications capabilities, and an information system interface based on Industry 4.0 concepts. Roadmap topics include:

- How communication and information management becomes more valuable as sensor data is integrated into business systems.
- How new technologies can simplify application engineering and maintenance by using smart, plug-and-play sensors.
- Required features and capabilities of smart process sensors — The roadmap includes a lengthy list of required smart-sensor capabilities. Such as autonomous, peer-to-peer sensor interaction, plug-and-play, self-configuration, and connectivity and communication using a unified protocol (OPC UA).

The process industry roadmap promotes a modular, minimum acceptable standards approach, which emphasises end-to-end system connectivity and data integration.

Digital manufacturing process systems connect applications, with which field devices communicate directly with process control, business, supply chain, engineering, and planning systems. Notice the difference between Industry 4.0 control systems and what came before. When processes become fully connected and integrated into a digital manufacturing framework, businesses can access all process-related data through networked systems. All the data that's gathered, analysed, and distributed applies to all stages of a product's life cycle. These expanded data management capabilities enable process industry companies to improve all their process- and business-related operations.

Making end-to-end connectivity work also requires system- and life-cycle-wide interoperability — significant obstacle to making Industry 4.0 work in discrete manufacturing. Interoperability poses challenges to process industry operations too.

14.3.3 Protocols & Minimum Acceptable Standards

In a fully interoperable, Industry 4.0 production system, all equipment, tools, and business applications can communicate without interference, data communication delays, or network downtime.

The ideal outcome of Industry 4.0 development is straightforward. Any machine that runs in a process environment transmits real-time details about its status and health, simply and easily. This data travels over any network and can be immediately understood by any other system that's based on the same operating standards.
In 2015, members of the NAMUR working group 2.6 (authors of the roadmap) recommended several Ethernet protocols to provide the minimum acceptable capabilities for the process industry.

14.3.4 Industry 4.0 in Process Environments

It's possible to build an Industry 4.0 process roadmap by using the high-level process described in Chapter 8, "Building an Industry 4.0 Roadmap that Works for Your Business." The process is straightforward:

- Start by considering the digital technologies and manufacturing ideas and performing the planning steps.
- Engage in a readiness assessment of any scope that helps you achieve your project's business goals.
- The core of Industry 4.0 process operation is a fully automated production line or facility, which would operate at ISA 95 Levels 0–2. Robust, well-designed readiness assessments include built-in IT and OT maturity analyses, which can help you identify the technology and process gaps.
- Your facility probably already includes sensors connected to pumps, motors, valves, and control systems as well as devices such as cameras and utility meters. Adding IIoT technology

will close existing gaps in data flow between production systems and between production and business systems.

- Consider adding raw material prices, social media product trend information, and many other types of mostly unstructured data to the now-business-wide data flow. The broader range of data supports more robust and comprehensive data models and promotes better data-based decisions.
- Consider using cloud-based services and technology, whether it's located off-premises or in a hybrid environment. Some SME manufacturers believe that they don't need off-premises data storage, at least not at the start of Industry 4.0 development. Although this might be true, cloud-based services promote high-speed, high-volume data management for data sets of any size. Making the data more accessible, especially when things go wrong.
- Cloud-based security services, for example, enable archiving, data scrubbing and other protection services. In a cloud-based service environment, data capacity and security become non-issues, two fewer things to worry about.
- Test your improvements in pilots that start small and can be improved quickly.

Finally, accept that you must stay in deployment mode, knowing that you never want to stop the process of continuous improvement.

14.3.5 Implementing Logistics 4.0 Practices

Today, companies wanting to survive in a hypercompetitive business environment must be willing and able to undergo profound changes in their technologies, processes, and business culture.

Logistics 4.0 is an industry-specific approach to using Industry 4.0 technologies and ideas. These technologies, processes, and practices extend basic strategies and process designs beyond the traditional boundaries of fabrication and assembly. As in Industry 4.0, these methods can optimise material flows and produce new business opportunities, in this case, throughout global supply chains— a familiar process.

In 2015, the early days of Industry 4.0 logistics studies, researchers found little interest in maturity models and logistics-related information. In "Intelligent Systems in Production Engineering and Maintenance," contributors Joanna Oleśków-Szłapka and Agnieszka Stachowiak present a framework of a Logistics 4.0 maturity model to fill the gaps.

A helpful approach to building Logistics 4.0 roadmaps uses the high-level methods of Industry 4.0 roadmaps used by discrete and process manufacturers. The process identifies major concepts and moves from planning, readiness assessments, and maturity models to detailed implementation and testing via an agile piloting process.

In Industry 4.0 research and development, maturity models provide an effective way to describe and measure the degree of progress made in each field of manufacturing. As usual with maturity models, the goal is to focus on current capabilities and provide detailed steps that can lead to continuous improvements in supply chain processes.

14.4 Two National Industry 4.0 Roadmaps

Since 2015, national and regional government agencies of the US, as well as ASEAN and EU member states, have launched documents that describe Industry 4.0 adoption processes. Unlike digital manufacturing frameworks, which describe national adoption resources and initiatives, roadmaps guide the adoption process with reviews of detailed procedures. We're providing examples from Singapore and Indonesia to compare the guidance supplied by Southeast Asian national governments.

14.4.1 Singapore's Industry Transformation Maps

Singapore developed its national Industry Transformation Programme as part of the nation's 2016 budget. The programme uses broad, sector-based strategies to adopt digital manufacturing technologies and methods.

In Singapore's Industry 4.0 schema, manufacturing is one of eight industry clusters. Sectors within manufacturing include precision modules and components, biologics and pharmaceuticals, energy, chemicals, marine and offshore, aerospace, medical-tech, and electronics.

The government started rolling out Industry Transformation Maps in 2016. These roadmaps guide digital manufacturing development and investment efforts, grouped by four key concepts: innovation; productivity; job redesign and skills improvement; and internationalisation of ITM strategies.

Here's how the programme's high-level development process works. The Economic Development Board of the Ministry of Trade and Industry:

- Rolls out sector-specific roadmaps for the 23 key sectors.
- Groups the sectors into six clusters to maximise collaboration opportunities.
- Finances Industry 4.0 adoption programs with initiatives outlined in national framework documents.
- Encourages representatives of trade associations, chambers, companies, and government agencies to work together to adopt Industry 4.0 technology, methods, and practices.

By using sector-specific industry transformation maps, Singapore's manufacturers can work from a government-approved framework. Take the aerospace sector, for example. The map uses the now-familiar approach of starting with critical concepts. In this case, these concepts include driving aerospace innovation via technology, increasing operational excellence, improving worker skills, and increasing collaboration among industry and cluster partners.

The framework also adds qualitative targets and several high-level KPI metrics. There are no detailed standards or performance recommendations. These are left to the discretion and experience of individual manufacturers.

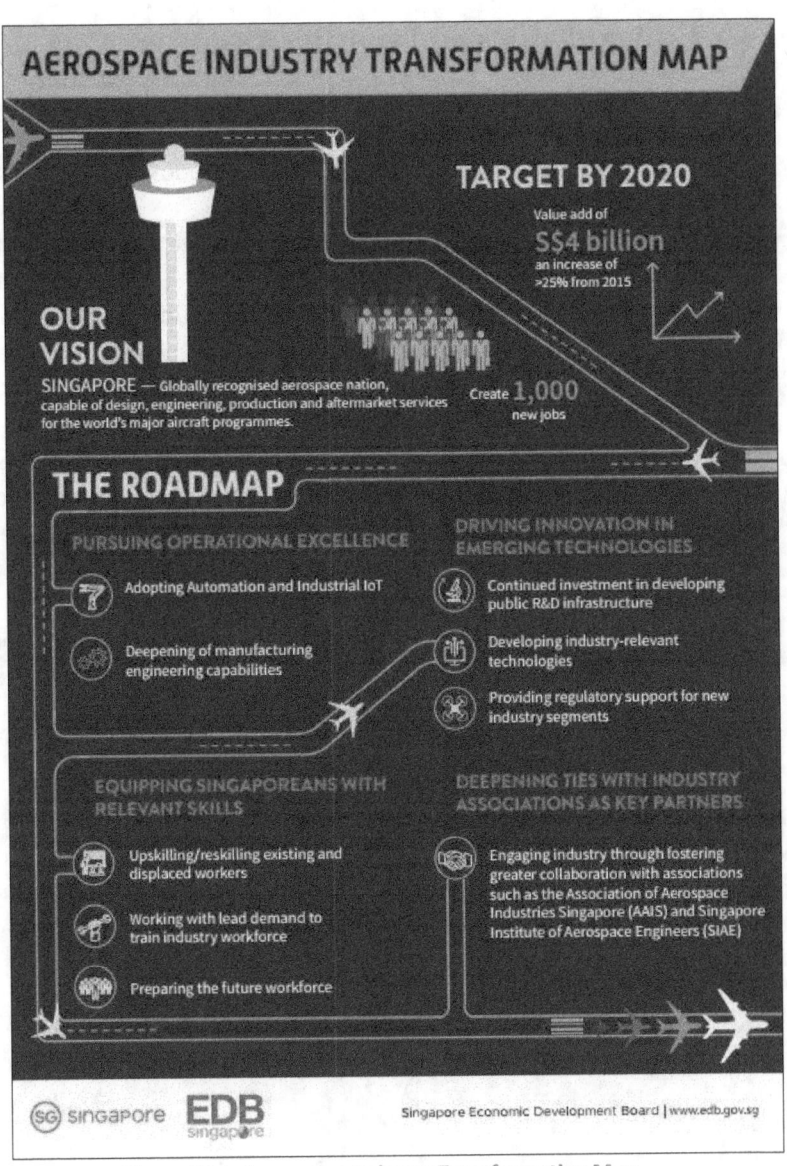

Figure 14-1: Aerospace Industry Transformation Map

(Source: Singapore Economic Development Board)

14.4.2 Indonesia's Policy Priorities Roadmap

Indonesia's digital manufacturing roadmap, Making Industry 4.0, uses a more general approach than Singapore's. Its provisions focus on high-priority sector development and government policies.

Much of the roadmap presents the case for digital manufacturing adoption among Indonesia's manufacturers. The document also identifies high-impact obstacles to this adoption and national economic growth. This information makes it a hybrid framework-roadmap document.

If there's a how-to aspect to this roadmap, it's indirect and focused entirely on high-level requirements. Success criteria are provided obliquely as lessons learned from other nations' Industry 4.0 journeys.

The guidelines focus on the government's high-priority manufacturing-related sectors. Then, the roadmap lists 10 national priorities. These policies describe what the Indonesian government can do to promote Industry 4.0 adoption.

Finally, Making Indonesia 4.0 concludes with goals. Some are aspirational. Others serve up measurable value points, such as GDP growth and job creation metrics with clearly defined timelines. It's not a meaty how-to document. However, its descriptions of policies, priorities, and high-level goals provide the beginnings of Industry 4.0 guidance.

15 Singapore's Industry 4.0 Story

"Every industrial revolution brings along a learning revolution."
- ***Alexander De Croo***

Our Industry 4.0 story has covered a lot of territory. From the ideas, visions, and realities that determine how new technology can redirect and invigorate manufacturing in good intentions, to hearing how digital transformation works in a real place — Singapore.

The Republic of Singapore is an ideal place to review and analyse early-stage digital transformation as it happens. Singapore has earned a reputation as a modern manufacturing hub. However, challenges within the country and from competing ASEAN nations test its leadership position. Government leaders look to Industry 4.0 technologies and ideas to make Singapore the most efficient, technologically advanced and competitive economy in the region.

A review of Singapore's print and online business press gives readers a good idea of the latest Industry 4.0-related topics and concerns. A quick read of the headlines provides the most prominent topics, such as:

- What's the status of manufacturing in Singapore today?
- What strengths and weaknesses will help or hinder the republic's development of its Industry 4.0-related goals?
- How might the government support and accelerate digital transformation?
- How might these capabilities and ideas enable a better way of life for Singapore's citizens?

277

The answers to these questions provide a clear way to monitor and analyse Singapore's status as a nation engaged in modernising its manufacturing sector.

15.1 Industry 4.0 Challenges & Opportunities

In 50 years of nation-building, there's no doubt that Singapore had developed a strong economy, one built on its manufacturing experience and know-how. All things change, however. Singapore's manufacturing sector is experiencing heavy competitive pressure from ASEAN rivals, global challengers, and the need to restructure its national economy. Long viewed as a key pillar of Singapore's growth, manufacturing now accounts for only a fifth of the nation's GDP.

The republic might have many advantages that its ASEAN peers lack (nearly universal internet connectivity, for example) but, government officials, educators, and business owners still confront social and economic problems that can stall Singapore's Industry 4.0 ambitions.

Beyond the global economic headwinds of the past decade, Singapore's manufacturers are fighting rising operations costs, more competitive ASEAN neighbours and a shrinking (and ageing) workforce. Given these stubborn obstacles to growth, government officials acknowledge the need to update its manufacturing model into one that offers more innovative, higher-value products and services.

15.1.1 Slow Pace to Adopt Ind. 4.0 In Manufacturing

Since Budget 2016, the Singapore government has introduced and funded many Industry 4.0-related financial grant programs for manufacturers. Nevertheless, the adoption rate of digital solutions remains low, especially among SMEs.

In the 2018 Association of Small and Medium Enterprises survey, only 57 per cent of SMEs polled were aware of digital transformation solutions at all. Of these respondents, 56 per cent of respondents had Industry 4.0 strategies in place.

In a similar 2018 survey by the Singapore Chinese Chamber of Commerce and Industry, 36 per cent of SMEs surveyed had not started to digitalise their businesses. Larger manufacturers did not do any better. In a recent survey by the Singapore Business Federation (SBF), only 36 per cent of businesses had consulted their sector's Industry Transformation Map to plan their digitalisation plans. While only 12 per cent of companies had invested in better training for their staff in 2018. This weak response has created concerns that Singapore's manufacturing workers wouldn't know how to use the technologies that their employers acquired.

What's going on? There are plenty of answers to the question, "What's slowing the adoption of Industry 4.0 in Singapore?"

- **Red tape** — Many SMEs cite compliance costs and slow processes as significant obstacles to rapid Industry 4.0 adoption.
- **Need for direct support on digital transformation** — More than 60 per cent of firms polled in the annual Singapore Business Federation's National Business Survey had a clear message: They want more government funding to support their adoption of Industry 4.0 technologies.

We've described strategies that help control some of the financial pain of adopting emerging technologies. However, some digital solutions can cost hundreds of thousands of dollars. So, despite government subsidies, adoption can swallow up a large portion of any SME's operating budget. This unwelcome reality makes Singapore's manufacturers look for even more government help.

- **Labour woes** — In the same SBF survey, 50 per cent of respondents cited rising labour costs as a problem. Forty-one per cent replied that finding workers with appropriate tech know-how was an obstacle. While another 29 per cent worried about providing their staff with digital skills.

- **A general economic funk** — Thirty-eight per cent of businesses surveyed in the SBF poll thought that Singapore's business climate would slump in 2019. Given that 88 per cent of survey respondents were SMEs, this pessimism strikes deep into the heart of Singapore's manufacturing sector. This is unlikely an environment that encourages risk-taking.

So, what do Singapore's government agencies, business owners, and manufacturing thought leaders propose to liven up the digital transformation process?

Getting things moving — Since September 2018, there have been calls for:

- **More public-private collaboration** — Singaporean manufacturers have made it plain that they look for more cooperation between the government and their industry. The desired goal: that the collaboration process channels available resources into digital transformation programmes. The desired targets are Singapore's manufacturers, especially SMEs.
- **Faster grant processes** — In a related topic, manufacturers not only want access to more grants and information that accelerate digital transformation. They want the government to quicken the pace of funding and information transfer. In response, the government has promoted the use of the Business Grant and SME online portals.
- **A faster path to solution commercialisation** — Singapore's manufacturers are gradually accepting the need to adopt Industry 4.0 technology and practices. However, they need solutions sooner rather than later. Faster commercialisation

requires faster communication of useful solutions between R&D specialists at Singapore's public research institutions and manufacturers. The goal is to accelerate the development of new products and services or new ways to streamline manufacturers' existing business operations.

The biggest challenge to Singapore's digital transformation, however, is the need for new jobs and skills.

15.1.2 The Need for New Job & Skills Training

"There is no room for complacency in the modern workplace," states Tristan Jinwei Chan in the South China Morning Post, "and Singapore should take note." With this clarion call to action, the article describes the need for Singapore's educators, business owners, workers, and corporate officers to step up and modernise their thinking about jobs and work skills. Building and maintaining a modern workforce requires flexible decision making, critical thinking, and a taste for lifelong learning. These habits are yet to be developed by Singaporeans, the author claims.

This article and others garnered from the nation's business press urge workers to master new skills ranging from using complex technology platforms to different ways of working and interacting with colleagues. The takeaway idea is that the country needs to redesign its jobs and upgrade employee soft skills to develop new ways of working together.

Paying more attention to the human element — Released in September 2018, the latest World Economic Forum Global Competitiveness Report reinforces the public declarations of concern. According to the WEF report, Singaporean companies and employees face major challenges of mental agility and the need to develop human resources in a world of digital transformation.

Singapore SMEs, which drive nearly 99 per cent of Singapore's economy, often favour the development of physical infrastructure rather than human capital. As a result, employers don't always see the need to invest in job and skills development, nor do they perceive well-trained employees as a competitive advantage.

Growing reliance on national funding — Many SMEs benefit from training grants and investment subsidies provided by government agencies. Programs such as SkillsFuture Singapore and Spring Singapore (renamed as Enterprise Singapore) are very popular. However, there's a growing body of evidence that unless training is subsidised, many SME manufacturers won't make training or skills development a high-priority business investment. For proof of this possibility, look no further than our survey quotes earlier in this chapter.

Need for employee retraining—There have been opinions on the headlines to encourage employees to take the reins of their careers in hand and show some personal initiative. This advice aligns with the government's roadmap shifting the entitlement mind-set of individuals and the need for individual employees to engage in self-directed learning and career-long motivation.

15.2 Industry 4.0 Strategies & Programmes

Do you want to improve adoption rates, readiness, or development of an Industry 4.0-related technology or human resources? If so, you'll probably find a committee, partnership programme, strategy, or collaboration framework to support it. Singapore's government and citizens have been diligent and energetic in establishing what amounts to nation-wide Industry 4.0 deployment tools. These groups and agencies are the fuel in the tank of Singapore's digital transformation.

15.2.1 National Labour Union Training Committees

During all the talk about Industry 4.0 readiness and roadmaps, the same question comes up again and again. Educators, workers, and business owners want to know the types of skills and training that workers will need to support digital manufacturing. The National Trades Union Congress (NTUC) has set up training committees, which are designed to do exactly that.

As of March 2019, the NTUC has worked with 10 companies in a pilot project to set up these committees. The partnerships include participation by about 64,000 workers in labour unions, industry firms, government agencies, and institutes of higher learning. The high-level goal of these groups is to help workers develop as their companies and industries change. After the pilot, the plan is to expand the collaboration to all NTUC unionised companies.

The training committees chart skills and gather competency requirements, put training programmes in place, and schedule worker training to minimise downtime caused by trainees going to class.

15.2.2 Smart Nation

The Smart Nation Singapore initiative is a national technology adoption programme, which promotes building a future society on a nation-wide technology infrastructure. Built on society-wide connectivity and adoption of converging technologies, this infrastructure is meant to improve Singaporeans' lives. By integrating technology into all aspects of citizens' lives, government officials hope to prepare citizens for future roles at work and in society.

Big ambitions for Smart Nation Singapore — In a May 2014 address to Parliament, then-President Tony Tan described the Republic's ambition to become the world's first Smart Nation. The president's vision was expansive. "We will make Singapore a Smart Nation: Enabling safer, cleaner and greener urban living, more transport options, better care for the elderly at home; more responsive public services and more opportunities for citizen engagement." In the five years since that speech, Smart Nation programmes and initiatives have proliferated and become familiar elements in leadership meetings and programme announcements throughout the country. Government officials have gotten mixed marks in periodic programme reviews in the print and online press. They also seem attentive to all the hot-button topics: cybersecurity, data governance, and infrastructure spending, for example.

Just another government programme? — One could view Smart Nation Singapore as one more visionary flight into the land of Industry 4.0. However, this view would be inconsistent with the persistent effort and funds that the government has put forward to make the Smart Nation concept work. The government seems to be very serious about the programme's success. However, it's time to change the focus from the government to the public. If the Smart Nation initiatives are going to work, citizens and business owners need to step up and show more enthusiasm about technology playing a bigger role in civil and business life.

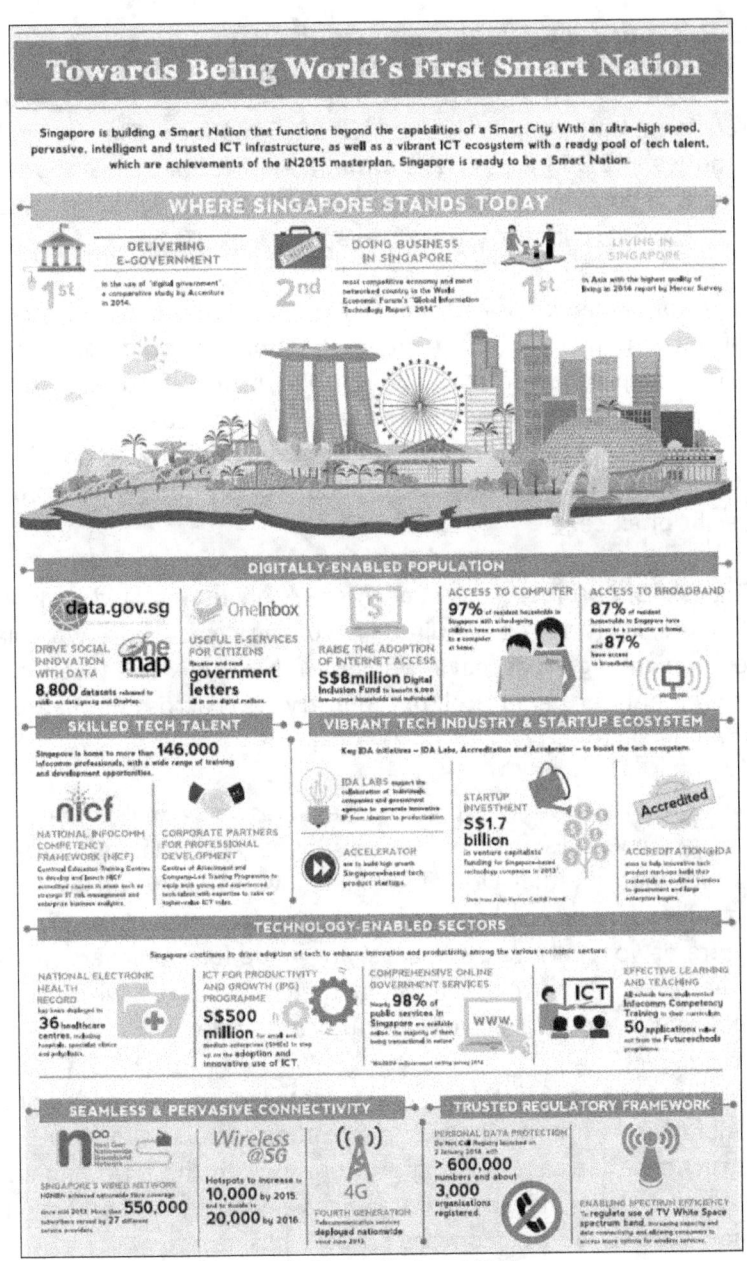

Figure 15-1: Singapore Smart Nation

source: http://smartisland.com/singapore-the-smart-island-smart-nation/

15.2.3 Industry Transformation Maps

The Singaporean government first announced the Industry Transformation Programme in its 2016 budget. The programme identifies and addresses transformation issues that are important to specific industries. A critical outcome of the programme was the development of Industry Transformation Maps (ITMs).

The government developed ITMs to look at industry-level transformation strategies in a more systematic and coordinated way. ITM development became necessary when manufacturers using the Singapore Industry Readiness Index (SSIRI) needed help. Without additional support documents and guidance, many companies had difficulty making the leap from Industry 4.0 theory to company-specific practice.

ITMs, skills, and jobs — Part of the ITM function is to identify which skills are likely to be required in current and future workplaces. Government agencies have worked with business, education, and government partners to roll out Industry ITMs across the nation's 23 economic sectors.

The Skills Frameworks Programme enables individual employees to identify new skills or improve existing ones in their industry (or another industry that workers might move to). Enterprises can also use skills frameworks to identify emerging job roles in their industry. The frameworks also provide a training guide for employers. With the frameworks, employers can equip their workforces with the skills employees need to fill current and future job roles.

ITM content and function — These detailed how-to guides provide manufacturers and businesses in 22 other sectors with step-by-step instructions that were missing previously. With the ITMs, companies could use their Industry 4.0 readiness data to create actionable roadmaps that were customised for their business.

Business owners customise the master manufacturing ITM to the needs of their business and Singapore's manufacturing industry. Government program designers reviewed the manufacturing landscape as well as future trends and requirement information. The information helps to establish a suite of initiatives designed to raise manufacturing productivity, develop employee skills, drive innovation, and promote internationalisation.

Each ITM consists of growth and competitiveness plans, which are supported by four sets of strategies:

- Productivity strategies, which help companies, especially SMEs, to move to higher-value-added activities and more efficient operations.
- Jobs skills modernisation, which includes investing in people, equipping them with deep skills and supporting the shift to greater value creation.
- Innovation strategies, which take advantage of Singapore's R&D resources to develop new products and services.
- Trade and internationalisation, which support companies that expand their business to overseas markets.

By February 2019, Singapore's government had launched ITMs for all 23 sectors of the nation's economy, but criticism of the roadmap process and results started long before then.

How useful are ITMs? — Each of the nation's 23 ITMs is tailored so that industries address specific issues through an approach that involves trade associations and chambers, firms, and the government. However, Ho Meng Kit, Chief Executive of the Singapore Business Federation, cited a big gap between the ITMs and the businesses they are designed to help.

Mr Ho also said small and medium-sized enterprises (SMEs) may find it difficult to understand the ITMs, which are "too high level"[3]. Ho also had concerns that ITMs were developed with larger enterprises in mind.

In newspaper articles and panel discussions, the usefulness of ITMs is still a matter of debate. Not all businesses are benefiting from industry transformation maps, and critics are not shy in outlining their deficiencies. The government's Economic Development Board started the ITM idea more than three years ago to help businesses propel the economy into the future. Since early 2018, critics have claimed that ITMs do not provide the information or capabilities that companies, especially SMEs, need to make their digital transformation, with the open and transparent information exchange from the stakeholders, ongoing effort are taken to fine-tune the framework and tools to ensure success of the transformation map.

15.2.4 Smart Factory Programme

If you wonder about what a smart factory is like, precision engineering company Feinmetall Singapore provides an excellent example.

A new factory — In June 2017, Feinmetall launched its 623 square-meters, S$6 million (US$4.5 million) manufacturing facility in Marsiling. The factory's advanced manufacturing features enable high-speed, high-volume machine data analytics. It uses IT and OT assets that minimise downtime, and plan machine maintenance schedules more effectively than before.

Feinmetall Singapore specialises in the design and manufacturing of wafer probe cards for semiconductor wafer tests. Almost total connectivity and interoperable IT systems throughout the new facility eliminate manual and paper-based tracking. The time-consuming, error-prone method of data analysis based on Excel spreadsheets is

---

[3] https://www.straitstimes.com/politics/industry-transformation-still-a-work-in-progress

gone for good. Now, data collection and analytics tasks are automated, like many other manufacturing processes in the factory.

The effort to improve human and production efficiency is part of the company's move from manual and low value-added work. Unlike some of its peers in Singapore, Feinmetall has set an explicit goal of reducing its human workforce without showing redundant operators the door. Facility managers want to upgrade operators into programmers, so they can earn higher salaries. The factory's human workers will need—and gets—higher-value training to work more closely with automated machines.

Feinmetall also plans to launch an e-portal to help clients troubleshoot and maintain the company's products remotely by using instructional videos. Company officials anticipate that this approach will reduce the need for its service engineers to make overseas trips to clients' offices to solve common problems.

Thanks to its investments in advanced manufacturing, Feinmetall is on track to meet its goal: doubling its annual revenue by 2020.

Smart Factory development options — Singapore's manufacturers have been slow to start their digital transformation process. However, interesting smart factory development strategies are emerging. Public and private programmes provide Singapore's manufacturers with a variety of paths to Industry 4.0 adoption.

Here are two approaches. The first describes the classic DIY method of building a digital infrastructure in-house with third-party solutions and tools. The other story describes a newer approach based on the collaboration of entities from Singapore's public and private sectors:

• **Get help from a high-value solution provider** — The first story describes one way to develop IT and OT infrastructure. Do it in-house and engage highly regarded engineers from an organisation such as the Singapore Institute of Manufacturing Technology (SIMTech). Then, have the experts guide you through the planning and installation process. SIMTech provides state-of-the-art tools and services for the analysis and optimisation of manufacturing process systems.

In October 2017, A*STAR launched its first SIMTech Model Factory. The precision engineering facility was the first initiative sponsored by the Industry Transformation Map programme. The factory enables SMEs to experiment with advanced manufacturing technologies in a real-life production environment.

• **Join a consortium to build a smart factory** — Public-private partnerships form the basis of the Advanced Remanufacturing and Technology Centre. The project is led by the Agency for Science, Technology and Research (A*STAR) and Nanyang Technological University.

ARTC is a consortium of more than 65 members, which range from SMEs to global multinationals. Focusing on manufacturing and remanufacturing the consortium speeds up the transfer of new ideas taken from applied research to commercial manufacturing applications.

15.2.5 The Model Factory @ ARTC

This model factory builds on Industry 4.0 principles and technologies. As part of the A*STAR's Model Factories Initiative, the program helps Singapore's manufacturers modernise their information and operations technologies. MF@ARTC focuses mostly on Industry 4.0 technologies, which can make industrial operations more efficient and competitive when used together.

Founded on an initial partnership of more than 20 industry and public sector research partners, the factory offers a state-of-the-art manufacturing facility equipped with industrial machines and digital systems throughout. Manufacturers, technology providers, and members of Singapore's manufacturing industry are welcome to visit the facility to learn how to use converging technologies for specific manufacturing use cases. The program also encourages testing of process improvements enabled by emerging technologies offsite, without disrupting their onsite operations.

In addition to the testbed, the factory also supports collaboration of manufacturers and their researchers. Their goal is to develop specific technology solutions for adoption at their industrial facilities. The factory offers capabilities and staff expertise in three areas: secure connectivity and intelligent systems, virtual manufacturing, and end-to-end solutions.

15.2.6 The National Robotics R&D Programme

In 2015, the Singapore government launched the National Robotics Programme to bring greater focus and consistency to the country's robotics-related R&D activities. In this programme, funds are available for research, living labs, testbeds, and projects that enable the mass adoption of robots. The programme scaled up its efforts in 2016 and announced more than $450 million to support its activities for the 2017 to 2019 period.

More recently, programme activities include supporting trade show, which displayed next-generation robotics products and solutions. Programme events reflect the NR2PO (National Research and Development Programme Office) support of solutions throughout the robotics value chain.

15.2.7 AI Singapore

This national programme focuses on the role of artificial intelligence and its ability to fuel the future digital economy. The programme is a nationwide partnership of all Singapore-based research institutions, key government agencies, and industry partners with AI expertise.

The AI Singapore programme is a government-wide partnership of the National Research Foundation, Smart Nation Singapore, the Digital

Government Office, the Economic Development Board, and many other agencies. Its goals, which include:

- Using AI to address major challenges that affect Singapore's society and industry.
- Investing in AI capabilities to exploit the next stage of scientific innovation. These innovations include next-generation, explainable AI systems, which have human-like learning abilities. Related technologies, such as computing architectures and cognitive science are also supported by the latest National Research Foundation schemes.
- Training local AI talent to increase their engagement with the development enabled by advanced AI capabilities.
- Broadening the adoption and use of AI and machine learning within the manufacturing industry. AI Singapore works with companies to use AI to raise productivity, create new products, and move solutions from the R&D lab to commercial products.

The National Research Foundation plans to deliver 100 meaningful AI projects and proofs-of-concept that solve real-world problems for end-users.

15.2.8 SkillsFuture Training Programme

Launched in 2014, the SkillsFuture movement focuses on Singapore's citizens getting the education, training, and lifelong learning opportunities that they need.

The programme provides citizens with a wide variety of learning opportunities. For example, a recent industry-relevant training program consisted of a list of short training sessions. Session topics focused on eight emerging and high-priority skills that are crucial to

Singapore's future prosperity and employment. By following programme guidelines, working adults can develop specific skills that meet changing job requirements in these eight areas. The goal: help Singapore's workers stay relevant and competitive in current and future job markets.

15.3 Service 4.0 Singapore

What is Service 4.0? — It is another Big Picture concept, the service analogue of products in the Industry 4.0 framework. Extending our attention to services is a very relevant idea for two reasons. First, Services 4.0 is a potential engine of growth for Singapore's digital economy. The nation's services industry accounts for 72 per cent of Singapore's GDP and 74 per cent of the nation's jobs.

Second, Services 4.0 was born from ideas in Singapore's Services and Digital Economy Technology Roadmap. Like manufacturing processes, services have evolved through the decades. Starting from relatively simple manual and internet-enabled forms, self-services came next, enabled by mobile, wireless, and cloud-based technologies.

If we are to believe IT marketers, the next phase in service delivery will be seamless, frictionless, empathic, and able to anticipate customer needs — enabled by emerging IT and OT technologies. Just as IT and OT infrastructures changed through time, services-related operations and opportunities will change, too.

In the Services 4.0 universe, there are two general approaches to delivering services in an "everything as a service" (XaaS) environment. Think of them as categories of value creation opportunities. They include:

- **Transforming the value/price equation** — In this approach, businesses choose to shorten their value chain, align prices with use, converge products, or unbundle products and services.

- **Harnessing network effects** — In this approach, businesses can expand their market reach, unlock assets from adjacent markets, turn products into product platforms, connect peers, and distribute service development resources and processes.

Services 4.0 creates a supporting framework of ideas and practices. With this approach, businesses can imagine, deliver, and manage future services. However, before the framework is useful, users must map their way through the framework. It's the only way the generalities of Services 4.0 can become useful — if you recall our roadmap chapter, mapping anything, be it technology or service, can be a slow, complex, and frustrating experience.

Services 4.0 programmes and initiatives — Singapore's government agencies are busy supporting programmes whose services can stimulate the use of convergent technologies. Here are a few of the ones most relevant to Services 4.0.

15.3.1 Cloud-Native Architecture

Since October 2016, Singapore's Info-communications Media Development Authority (IMDA) has regulated and stimulated the development of the republic's digital communications industries. IMDA initiatives support local info-communications technology (ICT) and info-communications media (ICM) industries.

The Technology Road Map identifies cloud-native architecture as the heart of evolving ICM ecosystems. The Cloud Native Architecture initiative makes it easier for local ICT companies to get access to emerging technologies. These are the enablers, which make digital communications more scalable, efficient, and cost-effective. IMDA also drives activities of the SG:Digital Cloud Community to promote the adoption of cloud-native architecture in Singapore.

15.3.2 GoCloud & GoSecure

This GoCloud capability development initiative helps SMEs in Singapore's local ICT sector keep pace with new technologies and industry practices. The programme uses cloud-native applications and microservices to improve the digital capabilities of SMEs.

GoSecure, another capability development initiative, helps SMEs in the ICT industry. The programme pairs SMEs with security specialists from the Singapore Institute of Technology to strengthen the cybersecurity capabilities of ICT companies.

15.3.3 Digital Services Laboratory

This facility brings engineers, local media talent, and other researchers together as a new research community. ICM (InfoComm and Media) users work with the laboratory staff to design and build solutions that address digitalisation challenges in several domain areas. The lab's speciality is developing solutions that provide faster knowledge and technology transfer. DSL projects also help to overcome barriers in large or complex projects.

15.4 Extracting Value from Industry 4.0

As development processes go, building Singapore's Industry 4.0 capabilities proceeds, but not as advertised. Real-life digital transformation in the Lion City is complex, unpredictable, and more challenging than expected.

Now that we've reviewed the rationale, enablers, and early results of Singapore's Industry 4.0 effort, it's time to return to answer the essential "so what?" question. Why bother developing Industry 4.0 technologies, use cases, roadmaps, and social infrastructure in the first place? The answer is straightforward: economic necessity.

In manufacturing, old-fashioned tactics, such as cost-cutting and low wages, aren't generating profits that they did in the past. Yes, "old-fashioned" efficiency improvements and cost reduction are still worth the effort of achieving. However, it's also time to review the opportunities provided by the revenue side of the equation. Creating more high-value products and services and nurture more satisfied customers, who are willing to pay for them.

So where is all this potential value going to come from? Start looking for it at:

- Connected processes that run throughout hyperconnected smart factories, offices, and warehouses.
- Expanded (and faster, more efficient) design, logistics, and after-sales processes.
- Innovations enabled by converging information and operations technologies.
- Focused government initiatives that inform business owners and fund tech adoption across the manufacturing sector.

Here are specific value threads that show the connections between convergent manufacturing technologies and value at the process, business, and national level.

15.5 What's next for Singapore?

Singapore's Industry 4.0 story is a mixed bag of impressive achievements and stubborn challenges.

The Republic's many infrastructure and business advantages put it well ahead of its ASEAN peers in terms of readiness. Recent developments provide examples of Singapore's ability and willingness to modernise its manufacturing environment.

Singapore continues to pour billions into becoming a testing ground of digital manufacturing innovation, with an emphasis on building IT start-ups and research. Part of the plan is for Singapore to construct a digital research district in northern Punggol. This site is where start-ups, established technology companies, and Singapore's Institute of Technology can work together and promote innovation in digital manufacturing.

16 Managing Change in Industry 4.0

"Two basic rules of life are: 1) Change is inevitable. 2) Everybody resists change."

- W. Edwards Deming

It's been said that successful adoption of Industry 4.0 technologies and methods is really an elaborate case of change management. Change management provides an excellent way to acknowledge the long-term value of smart manufacturing, and its gritty real-life challenges.

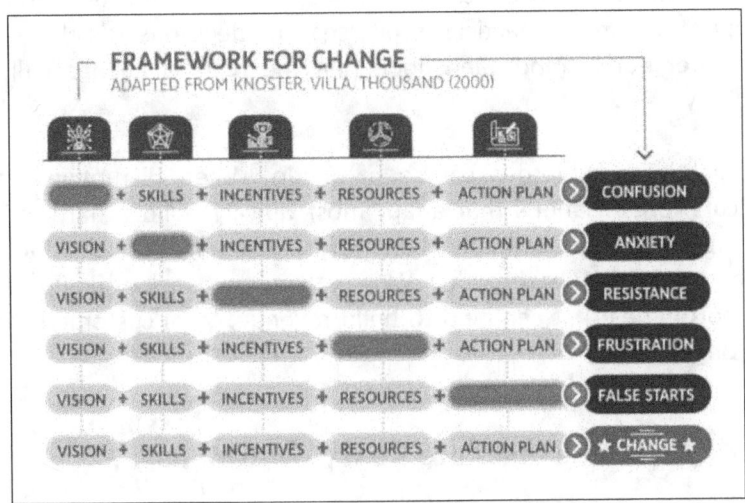

Figure 16-1: Framework for Managing Change

In plain terms, change management involves the use of every tool available—usually, technology, funding, and the power of human persuasion—to achieve desired business results. You can mention redirecting resources, changing business processes, or rerouting budget allocations. However, ultimately, you want to use resources to move your organisation from a beginning to a final state.

16.1 Change Management in Industry 4.0

Managing change in smart manufacturing environment involves surviving global shifts in manufacturing, logistics, and related services, which occur at an increasing pace. Globalisation and constant technological innovation create a roiling atmosphere in a constantly evolving business environment.

With local and global business environments undergoing so much change, organisations can learn to become comfortable with change itself. Yet, every business owner or leader knows that organisational change at any pace is challenging. The structure, culture, and routines of organisations often create a persistent and difficult-to-remove imprint of familiar practices, processes, and decisions. It's all too easy to enter denial mode and hope that the rush of change will pass quickly.

Denial is a dangerous course of action. When disruptive change occurs, organisations that adapt most quickly build a hard-to-beat, competitive edge. Companies that refuse to change lose the chance to create efficiency-based value. Worse, they don't seek or find potential opportunities that can help to build revenue, loyal customers, and a stronger brand.

16.1.1 What Changes are we trying to Manage?

Conventional definitions of Industry 4.0 change management focus on improving efficiency and building long-term value by changing manufacturing technology and methods. Everything is couched in terms of choosing, adopting, and orchestrating technology to generate value. The goal is to search for, pursue, and benefit from new value growth opportunities enabled by converging manufacturing technology.

Ultimately, the change that's managed in Industry 4.0 environments is human behaviour. Human collaboration, habits, and mind-sets take the form of actions made at the levels of national policy, business strategies, manufacturing processes, and individual decisions.

At the national level, smart manufacturing changes involve political action taken to define relevant policies, create government programs, and support Industry 4.0-related educational goals and achievements.

Business-level changes reflect group decisions made by company business and technical leaders. Process-level changes involve the design and operation of manufacturing products, processes, and services within smart factories, warehouses, and offices.

Finally, behavioural change at the individual level includes making educational and training choices, developing a modern work mind-set, and accepting work styles, such as collaboration and life-long learning.

16.1.2 Accelerators & Obstacles

Identifying accelerators of smart manufacturing adoption and obstacles to it is a useful way to analyse patterns of needed change. These indicators make it easier to identify and (potentially) change what can be improved in smart manufacturing operations.

Government documents, print and digital news outlets and countless online business commentators throughout ASEAN list and describe obstacles to Industry 4.0 adoption. These are the challenges that manufacturers face when they engage with smart manufacturing technology and ideas.

Smart manufacturing accelerators acknowledge a problem and often offer improvements or enable change. Accelerators can be a grant funding program for SMEs, an international partnership of educators and businesses, or a vocational training program that sends trainers to factories. These resources and relationships enable adopters to build value more quickly and with fewer resources than traditional manufacturing technology and methods require.

As you might expect, obstacles and accelerators number in the hundreds. When you group them in terms of our four levels, interesting and useful patterns emerge.

16.2 Industry 4.0 Change Management in ASEAN

Compiling an accurate, up-to-date account of Industry 4.0-related challenges and enablers is a good way to identify changes that business owners must make to benefit from smart manufacturing technologies and practices.

Taken from ASEAN-6 news outlets and government documents, this information acknowledges the need for specific types of change in different Industry 4.0 environments. To look for useful patterns, we grouped changes at the four levels described earlier. This is what we found.

16.2.1 National Level

We're starting with changes at the highest level, that of government policy and decision making. These are the most familiar accelerators and obstacles, which are often mentioned in the news and speeches made by national politicians.

Government policy — What can ASEAN's national policy-makers do to promote or speed up Industry 4.0 adoption and value building among manufacturers? They develop and support programs and initiatives, lots of them. These initiatives are designed to strengthen awareness and acceptance of smart manufacturing value and ideas. They use each nation's policy apparatus to lower barriers to entry of modern manufacturing and align workforce resources with employer requirements.

Readiness assessments — Governments such as Vietnam, Singapore, and Thailand are creating national roadmaps to adopting smart manufacturing technologies. Singapore's government developed and promoted the Singapore Smart Industry Readiness Index, a first-of-its-kind self-assessment tool. Designed for use by Singapore's manufactures, the tool's usefulness is extended by newly developed software, the Smart Industry Readiness Prioritisation Matrix

Regulatory, business, and trade policy — Many Industry 4.0 programs throughout ASEAN try to reduce the effort and cost of producing and selling manufactured goods within and beyond ASEAN. These initiatives include measures such as defining new tax rules and trade regulations to promote the immediate start of digital transformation. Other measures address manufacturing-related tax reform streamlined regulatory policy and changing trade policies that hurt advanced manufacturing firms.

Funding — Programs and initiatives that provide inducements for manufacturers to adopt Industry 4.0 principles are a big part of promoting smart manufacturing. ASEAN governments have heard the call of the region's manufacturers for access to more grants and information that can accelerate digital transformation. They have built and promoted special programs such as the Business Grant and SME portal programs run by the Republic of Singapore.

Funding and promoting national education and training programs also play large roles in national Industry 4.0 policy. These funds include everything from forming partnerships with German manufacturers with apprenticeship programs to three-hour workshops for manufacturing employees who want to upgrade their skills.

Partnerships — News about ASEAN national governments, their educational institutions, and manufacturers communicating and working together takes up a lot of space in print and online publications. Partnerships of all kinds are popping up within individual countries and throughout the region. New partnerships between educational institutions and ASEAN manufacturers, industry-university groups that collaborate in advanced manufacturing research, and groups that identify obstacles to digital transformation are in the news.

In these groups, leaders from the public and private sectors work together to identify specific barriers to smart manufacturing and the opportunities that Industry 4.0 ideas enable. ASEAN-wide exchange and cooperation are promoted by pilot and demonstration projects sponsored by regional programs and partnerships.

Education and training — In ASEAN countries generally, there aren't enough properly trained operators, technicians, and technical professionals to fill the jobs that manufacturers need onsite today. In countries such as Singapore, an ageing, shrinking workforce makes the job vacancy problem even worse. As a result, government-level efforts to promote better job-worker alignment are in progress.

Nationwide, skills development projects — These programs, such as Singapore's SkillsFuture initiative, are lifelong skills development programs. Guided by the republic's Future Economy Council, this initiative brings together talent from government ministries, corporations, academic institutions, and labour groups. Their goal is to design curricula that can fill current job openings and help workers at all levels keep their knowledge and skills up to date.

Training programs that teach skills those local manufacturers need. Getting the right manufacturing skills and training to workers in manufacturing hubs is a serious challenge to ASEAN educators. That's why education and training institutions are starting to change vocational education to make students job-ready for factory work.

Programs that develop more technical managers — Manufacturers, especially larger enterprises, need employees who can manage workers who operate advanced manufacturing technology. This requirement translates into a need for more engineers and technical experts with backgrounds in electronics, material science, and medicine.

Development of modern teaching and learning practices — Lecture- and memorisation -based teaching methods favoured by East Asian schools have been a staple of ASEAN education systems. This strategy no longer produces graduates with the skills needed to operate and manage smart manufacturing facilities. Now, national government and educational institutions are paying more attention to blended, practice-oriented, project-based, and scenario-based learning.

The old "Graduate and you're done" approach to education is fading. Now, educators are starting to emphasise lifelong learning in curricula and post-graduate programs. Lifelong learning is slowly becoming a permanent part of professional life.

Looking for and following successful examples — Other countries, such as Germany, have addressed and managed technology and labour challenges that ASEAN educators face today. Some ASEAN programs highlight the following successful educational models. These programs combine formal learning, vocational training, and practical work experience, which resemble the European apprenticeship system.

Product commercialisation — The inability to commercialise a manufactured product with local talent and resources creates a gap in ASEAN manufacturing value chains.

Creating a supportive environment for innovators and entrepreneurs requires R&D infrastructures and public-private support networks. Japan, Germany, and the United States required decades to build these essential institutions. ASEAN governments hope that their countries can use advanced technology and newly formed partnerships to leapfrog over these difficulties. However, likely achieving these goals might be, ASEAN governments offer many approaches to building manufacturing R&D infrastructures. These methods include:

- **Developing commercialisation strategies** — One method proposes a national advanced manufacturing strategy. This approach involves setting up a systematic process, which

305

identifies and prioritises each nation's cross-cutting technologies.

- **Building innovation infrastructures** — Maximising the value of smart manufacturing technologies requires the knack for innovation. Relatively few manufacturers throughout ASEAN have the entrepreneurial skills needed to innovate. This lack of infrastructure for technical and business innovation are the serious barriers to Industry 4.0 adoption and value optimisation.

- **Increasing R&D funding** — This approach establishes a starter list of cross-cutting technologies that are vital to advanced manufacturing. It also uses the U.S. Advanced Manufacturing Partnership process as a template to evaluate technologies for R&D funding.

- **Support training and talent initiative** — ASEAN governments also encourage pilot and demonstration projects to develop high-level design and engineering talent.

- **Advanced tools for advanced manufacturing** — Governments also established a national advanced manufacturing portal. These programs create searchable databases of manufacturing resources that provide SMEs with a support infrastructure.

- **Less bureaucracy, more help, please** — Despite government efforts to speed smart manufacturing innovation along, there will always be something that makes manufacturers reluctant to engage with Industry 4.0 programs. For example, difficulties in handling the paperwork and documentation are a perpetual problem. Malaysian manufacturers, for example, have a hard time understanding whether they qualify for various programs. Government websites and documents often provide little help, which manufacturers need in getting the application process started. They would also like government agencies to provide more precise guidelines about joining national initiatives and programs.

16.2.2 Business-Level Change

At the level of individual businesses, obstacles and enablers operate though decisions made by different manufacturers in ASEAN-6 nations. From the top to bottom of each manufacturing business, Industry 4.0 technologies and methods are making fundamental changes in how people work. These changes have the biggest impact on the mind-set, expectations, habits, and pay checks of each company's leaders and workers.

Most ASEAN manufacturers know that changes can improve process efficiency and cost control and grow long-term business opportunity and value. However, many manufacturers are slow to accept the need for new behaviours and outlooks, as well as technologies and methods.

Concerns about the uncertainties of digital transformation take several forms. They can appear as low Industry 4.0 adoption rates or assigning low priorities to discover new business models to build long-term value.

Near-universal unease about security — Well-founded worries about hackers and cyber attackers have been a major barrier to adopting smart manufacturing methods. This trend is especially true among ASEAN's SMEs. Manufacturers view the all-connectedness of Industry 4.0 environments with mixed feelings. They recognise the advantages of connectivity and the vulnerability that it creates for their data and websites.

Use of automation and AI capabilities — Deciding how much automation to use on the shop floor will be up to the technical and business leaders in each manufacturing business. Managing the inevitable shift in workplace dynamics is an often-overlooked topic of

307

discussions of Industry 4.0 adoption and management. Manufacturing leaders must decide whether, how, and how much that humans work with machines.

Managers and executives recognise the value of automating at least some tasks by machines. However, dealing with the changes in the work environment requires a new and different approach to workplace dynamics.

Digital transformation costs — All manufacturers keep a close eye on their technology budgets. Nevertheless, in surveys, SMEs often view the costs of adopting flexible automation technology and processes as a significant challenge.

Some of this concern is unnecessary, however. Although technology costs can crush an SME's annual technology budget, cost-cutting workarounds are available. For example, a common misconception is that adopting automation technology is always expensive. This idea creates real problems, but it's based on the assumption that manufacturers must automate entire buildings at one time. Instead, manufacturers can design and run a pilot on a single production line onsite or in a government-sponsored sandbox environment.

When used in an agile production environment, companies can develop small-scale process improvements quickly and at far less cost than all-at-once deployments. The details of setting up small-scale pilots belong in the discussion of process-level changes. However, the decision to run these pilots at all belongs with company decision-makers.

Changing customer tastes and expectations — The combined capabilities of semi-autonomous machines, versatile robotics designs, and faster, more flexible human-machine workflow open new opportunities for agile manufacturing.

These enablers make it possible to offer customised products and faster, on-time delivery. This is a good thing because customers are developing a taste for personalised goods and rapid delivery. Using intelligent, versatile, and highly integrated systems plays an essential role for manufacturers who give high priority to customer service.

Modernising business infrastructure — Many ASEAN countries are still in the process of buying and integrating connectivity, information, and communications technologies of earlier times into their business and manufacturing infrastructures. Modernisation brings its own change requirements.

Playing catch-up throughout the region — Some ASEAN manufacturers score low Industry 4.0 readiness because they still use machines, devices, tools, and methods from earlier industrial revolutions. This lack of preparedness is largely due to unequal industrial development in different regions of ASEAN member countries. Manufacturers won't be able to adopt Industry 4.0 technologies until they modernise their business communications and connectivity infrastructures.

Standardisation of advanced technologies and methods — This process has hardly begun in ASEAN nations. Everyone recognises the need for new open standards, which help support the seamless interchange of data across systems but follow-through has been slow. These standards are essential to controlling and protecting signals, sensor measurements, and other data that's continuously exchanged in real-time.

A more modern view of logistics and supply chain management — Turning massive volumes of IIoT data into actionable intelligence is key to smart supply chain management in smart manufacturing operations. Before that can happen, however, it's necessary to bring logistics and SCM under the umbrella term of "production." Like design and after-sales services, logistics takes the production concept far beyond the traditional notion of production as fabrication and assembly alone.

Modernising business culture — Successful digital manufacturing companies would benefit from a more modern approach to business culture and the many relationships that occur within and beyond the smart factory. These company-wide changes reflect new priorities in production operations, budget allocation, and hiring practices.

Paying attention to training and HR policy are good examples of this change. Thriving Industry 4.0 programs are more likely to build long-term value when they include more investment in people and change management. Recent surveys suggest that many ASEAN manufacturers assign lower importance to modern employee education and training. These respondents also undervalue employees with this training as a competitive asset.

Expanding the view of smart manufacturing benefits — Manufacturers in countries throughout ASEAN recognise and pursue immediate value—more efficient production and more productive employees. However, using smart manufacturing technologies and methods to develop new, lucrative business models has a much lower priority on executives' radar.

Giving customers a more significant role — Several trends, dealing with what one could categorise as "improved customer consciousness," have come up in the United States and European Union.

One of these developments involves getting a broader, more immediate view of product development from the standpoint of customer stakeholders. In aerospace, automotive, and other large-scale manufacturing endeavours, production cycles are long and rework costs can be considerable. Increasingly, enterprises are making more customer stakeholders part of their design and production processes.

This approach gives technical and business decision-makers an opportunity. Manufacturers can extend the connectivity that links their production operations and business decision-makers to a larger group of customer stakeholders. The goal would be getting more informed opinions from clients to reduce the risk of redesigning or reworking products.

Making collaboration and communication a habit — As the stakeholder example above shows, improving connectivity does much more than knocks down data silos. Changing attitudes and strengthening relationships with teams throughout the factory and business are essential, too.

Smart manufacturing merges IT, OT, and business assets and connects them to a local network and the internet. This arrangement puts joint responsibility of all formerly segregated departments to communicate and reach consensus on a wide range of activities and policies.

Getting rid of the Industry 4.0 mystique — Finally, manufacturers won't optimise value via digital manufacturing unless all employees understand what Industry 4.0 means to each manufacturer's business goals and employees' jobs.

Business leaders must educate machine operators, technicians, engineers, and other production professionals as to why and how smart manufacturing matters to them and the company. Emphasising the business case and workforce benefits of moving to a smart manufacturing environment will go a long way to achieve employee buy-in.

16.2.3 Process-Level Change

Most change management discussions focus on what happens at the process level. Designed by engineers and monitored by managers in specific departments, these changes involve choosing, configuring, and operating various technologies at different parts of a smart manufacturing facility. Process changes occur in smart factories, warehouses, or offices. Changes such as the choice of specific technologies and the pace of their adoption made at this level will play a fundamental role in building long-term value.

Hundreds of processes have roles in each smart manufacturing facility and office. The process changes described here reflect only primary considerations, such as infrastructure improvements, process planning, and system control. Information, operations, and mechanical infrastructures merge in smart manufacturing processes.

Data and security infrastructures — Information infrastructures connect mechanical and IT assets with IIoT sensor devices, generating high volumes of data in high-speed operations. Legacy data infrastructures usually run on low-bandwidth connections that are not suited to smart manufacturing environments. As a result, network stability and reliability are common problems, which will need correcting.

Data collected and information created in a data infrastructure must be secure when it's in transit and storage. Securing a hyper-connected manufacturing system within and beyond the smart factory often requires plenty of time, effort and a hefty budget. Different security services and system configurations are available to many ASEAN manufacturers now.

System interoperability — The relatively low level of manufacturing system interoperability throughout ASEAN is a startling finding. Machine-to-machine communications and many other smart manufacturing processes depend on end-to-end connectivity. This, in turn, requires interoperability throughout the life cycles of all connected systems.

The lack of interoperability is a major obstacle to smart manufacturing in discrete manufacturing and process industry operations. The need to integrate manufacturing systems with the internet adds a layer of complexity, which is likely to increase the time and costs of IIoT deployments. Acceptance of technical standards for operations and communications processes is the answer to this problem. However, it's difficult to predict how or how long it will take for standards-setting collaboration to take effect.

Asset tracking technology — Smart manufacturing methods encourage a broader view of production processes. This definition includes the IIoT-connected supply chain. Internet-connected sensors are applied to production assets, products, and the vehicles that move them. Development of less complex and expensive IIoT devices and systems makes real-time asset tracking accessible to ever-smaller companies. New asset management software also makes tracking data more visible.

Better version control capabilities are also available. Even within one company, it can be challenging to ensure that the design engineers and the machinists are looking at the same drawing. However, expand that problem across an entire supply chain, and version control becomes a vital application area that smart manufacturing technology must address.

Finally, Industry 4.0 practitioners see the criticality of managing change at the human level.

16.2.4 Individual-Level Change

Eventually, change management considerations come down to human behaviour. At the level of individual decision-making, these changes reflect how people choose their education or training or learn to collaborate with business peers, partners, customers, and robots.

Education and training — Much of individual-level change management involves learning. This can be formal education or training or include informal advice or instruction from mentors or peers. Whatever the learning environment, however, the outcome is growing knowledge, skills, and experience.

Robots and the human element in smart manufacturing — Smart manufacturing is based on automation and decentralised decision making. To satisfy the requirements of decentralisation, it's essential to set up and maintain an automated, intelligent, and increasingly autonomous flow of assets, goods, materials and information.

However, no matter how autonomous we want systems to be, an important human element remains. For now, at least, smart manufacturing needs people to plan and act because not all actions can be automated.

Awareness of Industry 4.0 value — Perhaps the most challenging obstacle to smart manufacturing success in ASEAN is the inherent conservatism of many manufacturers. Survey results provide a worrying indicator that manufacturers aren't aware of or don't believe that there's plenty of value to be an early Industry 4.0 adopter.

Companies might be waiting for the price of smart manufacturing technology to fall. Or, they might be waiting for early adopters to succeed or fail. Either way, waiting might be an expensive choice.

Modernising the business mind-set — One of the more fascinating aspects of change management is that it's not just about technology. It's also about human mind-sets, decision-making styles, and how people can gain, or lose, competitive advantage and opportunities for long-term value growth.

Much of the value of smart manufacturing technology will belong to innovators and entrepreneurs, who convert new ideas into manufactured goods and related services. Legacy technical and business thinking will make it difficult and expensive for traditional manufacturers to compete.

No matter what type of job shortage we mention, fixing the shortage must wait until individuals take the plunge into a new or newly changed manufacturing career.

The graphical illustration below provides a wider horizon for planning business architecture with consideration of technologies application and end-to-end business environment in various verticals and segments.

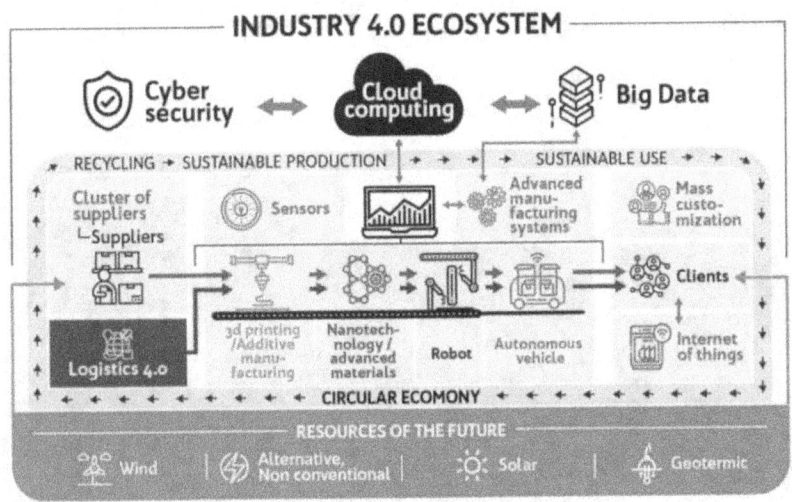

Figure 16-2: Industry 4.0 Ecosystem

17 ASEAN's Industry 4.0 Journey

"If you can't fly, then run, if you can't run, then walk, if you can't walk, then crawl, but by all means keep moving."

- Martin Luther King Jr.

We've covered a lot of territory since we started our Industry 4.0 explorations on the VinFast manufacturing floor. We profiled the latest use cases of robotics, augmented reality, data analytics, and other technologies that are becoming mainstays of modern manufacturing. We tracked incentives supported by ASEAN-6 national governments and the beginnings of region-wide, multinational partnerships. We also reviewed the development of the Vietnamese software market, Thai robotics, and Singapore's innovation ecosystem.

These are just a few of the many examples of Industry 4.0-related technologies, manufacturing processes, and practices taking hold within ASEAN-6 nations. In closing, we provide two concepts, which you can take away and use in your engagement with Industry 4.0.

17.1 Broader View of Manufacturing Technology

The development and adoption of new technologies that knockdown data silos encourage a new view of manufacturing workflow.

Until very recently, "manufacturing" included the resources and processes involved in fabricating and assembling parts into finished products. Industry 4.0 concepts such as smart factory design and the RAMI 4.0 model expand this idea. Big data analytics, machine learning, and ICT enable manufacturers to connect a massive variety of cyber-physical systems and analyse the data they generate.

This, in turn, helps manufacturers expand workflow into a series of steps that range from design to post-sales services. By enlarging the definition of manufacturing, Industry 4.0 workflow occurs in design works and smart offices as well as on the shop floor. Ultimately, value derived from manufacturing can be discovered in these new areas too.

17.2 Playing the Long Game

The success that ASEAN-6 nations achieve will depend in large part on their ability to take a strategic view and deal with change.

Obstacles and accelerators, opportunities and challenges, these words appear again and again throughout this book because they reflect the ever-present potential of building business value by harnessing Industry 4.0 technologies and ideas. Making the most of Industry 4.0 resources and ideas requires a willingness to develop the habits of long-term strategy and planning.

The public face of this long game involves the creation of government policy, regulations, and investments in education and support infrastructure such as ICT. National governments that clearly define Industry 4.0 policy goals and support manufacturers in down-to-earth ways are developing the beginnings of smart manufacturing capabilities.

The private side of Industry 4.0 strategy also requires a clear vision of long-term goals that is individual businesses looking ahead. They need a guide that assesses their technology maturity and a roadmap that can help them prioritise and implement tactical decisions. The goal: gradually upgrading their production and business capabilities and perhaps supporting them with value-added services.

17.3 Recapturing ASEAN's Manufacturing

It's possible to revitalise the manufacturing capabilities of former ASEAN manufacturing powerhouses and nourish the resources of newcomers.

ASEAN used to be a global manufacturing hub. Regaining this status has been a big topic of media outlets within and between ASEAN member countries. However, regaining that position demands a daunting list of changes, which will be easier or more difficult, depending on the country. The changes include demands made on:

- **Governments** — Providing clearly defined manufacturing policy goals and offering convincing financial and educational incentives.
- **Educational institutions, training businesses, and partnerships** — Investing in education, training, and retraining that provide ASEAN nations with badly needed manufacturing knowledge and skills.
- **Individuals** — Convincing the region's students to consider study in technical and manufacturing-related business skills.
- **Business owners and executives** — Becoming more risk-tolerant in strategic decision making and more change-tolerant in general would be an excellent start. Then, they might also reconsider the value of independent-minded employees and new revenue sources provided by new Industry 4.0-enabled business models.

Expansion, Opportunity, Change... these are the bywords of improving economic prospects that Industry 4.0 ideas enable throughout ASEAN. The trick, of course, is to muster the individual and collective courage.

References

Agarwal, A., Lath, V. and Mancini, M. (2017November), Confronting Indonesia's Productivity Challenge [White paper], Retrieved from https://www.mckinsey.com/featured-insights/asia-pacific/confronting-indonesia-productivity-challenge.

Alan N. Daum and Associates (2018 December 20), OPC: Interoperability standard for industrial automation [Blog post], Retrieved from https://www.adaum.com/2018/12/20/opc-interoperability-standard-for-industrial-automation.

Alton, C. (2017 August 19), Differences Between Field, Control, Supervisory, and Enterprise Levels of Automation [Blog post], Retrieved from https://learntechnique.com/differences-field-control-supervisory-enterprise-levels-automation.

Andrew Ross. The Information Age. Retrieved from https://www.information-age.com/5g-is-the-heart-of-industry-4-0-123483152/

Anunziata, M. (2018 August 10), Seven Steps To Success (Or Failure) For 'Made In China 2025' [Online article], Retrieved from https://www.forbes.com/sites/marcoannunziata/2018/08/10/seven-steps-to-success-or-failure-for-made-in-china-2025/#4aa388ea4057.

Arbulu, I. et. al. (2018 February) Industry 4.0: Reinvigorating ASEAN manufacturing for the future [White paper], Retrieved from https://www.mckinsey.com/business-functions/operations/our-insights/industry-4-0-reinvigorating-asean-manufacturing-for-the-future.

Ariffin, E. (2019 February 26), Can Industry 4.0 revolutionise manufacturing? [Blog post] Retrieved from https://theaseanpost.com/article/can-industry-40-revolutionise-manufacturing.

Asian Robotics Revuew staff (© 2019) Thailand's $45 Billion Leap to Industry 4.0 [Blog post], Retrieved from https://asianroboticsreview.com/home92-html.

A.T. Kearney (2018 October), Making Indonesia 4.0 [White paper], Retrieved from https://www.bu.edu/alumni/files/2018/11/AF18_Shirley-Santosa.pdf.

AT Kearney. Accelerating 4IR in ASEAN: An Action Plan for Manufacturers. Retrieved from https://www.atkearney.es/operations-performance-transformation/article?/a/accelerating-4ir-in-asean-an-action-plan-for-manufacturers
Austrian Office of Science and Technology (© 2019), Industry 4.0: Manufacturing in the United States [Online magazine article], Retrieved from https://www.ostaustria.org/bridges-magazine/item/8310-industry-4-0.

Aziz, A. (2018 October 31), Malaysian PM: Industry 4.0 adoption 'boils down to knowledge on application'   [Blog post], Retrieved from https://www.theedgemarkets.com/article/malaysias-industry-40-adoption-boils-down-knowledge-application-says-dr-m.

Basl, J. and Doucek, P. (2019 March), Metamodel for Evaluating Enterprise Readiness in the Context of Industry 4.0 [Research article], Retrieved from https://www.mdpi.com/2078-2489/10/3/89/htm.

BDO USA (© 2018), The Middle Market Manufacturer's Roadmap to Industry 4.0 [White paper], Retrieved from https://www.bdo.com/getattachment/3f03fd57-6498-4bf1-affe-cddb0b13d6ed/attachment.aspx?2017-M-D-Industry-4-0-Special-Report_WEB.pdf.

Bhunia, P. (2018 January 14), Only 25 countries well-positioned to benefit from Industry 4.0 according to new World Economic Forum report [Blog post], Retrieved from https://www.opengovasia.com/only-25-countries-well-positioned-to-benefit-from-industry-4-0-according-to-new-world-economic-forum-report.

Bhunia, P. (2018 March 20) Three strategies for Thailand's economic transformation [Blog post], Retrieved from https://www.opengovasia.com/three-strategies-for-thailands-economic-transformation-mega-projects-eastern-economic-corridor-digital-development.

Byrne, A. (2018 July 30), Why ambitious Industry 4.0 plans need a better open standard [Blog post], Retrieved from https://www.iottechnews.com/news/2018/jul/30/why-ambitious-industry-40-plans-need-better-open-standard-approach-succeed.

Byrne, A. (2018 July 14) Why Industry 4.0 will require us to change the way we think of managing Big Data [Blog post], Retrieved from https://www.manufacturingglobal.com/technology/why-industry-40-will-require-us-change-way-we-think-managing-big-data?utm_content=74357697&utm_medium=social&utm_source=twitter.

Chan, E. (2018 September 10), Made in China 2025': is Beijing's plan for hi-tech dominance as big a threat as the West thinks it is? [Newspaper article], Retrieved from https://www.scmp.com/business/china-business/article/2163601/made-china-2025-beijings-plan-hi-tech-dominance-big-threat.

Chan, T.J. (2018 April 26), What Industry 4.0 means to Singapore and why its workers must upskill and lose their sense of entitlement [Blog post], Retrieved from https://www.scmp.com/lifestyle/article/2143239/what-industry-40-means-singapore-and-why-its-workers-must-upskill-and-lose.

Chang, K. (2017 December 26), What Singapore's Smart Industry Index Can Tell Us About Industry 4.0 in 2018 [Blog post], Retrieved from https://blog.inmindcloud.com/singapore-smart-industry4-manufacturing-2018.

Chatterjee, R., and Gamota, D. (2019 July 2), Smart Manufacturing Roadmap [Blog post], Retrieved from http://smt.iconnect007.com/index.php/article/118153/smart-manufacturing-roadmap-data-flow-considerations-for-the-electronics-manufacturing-industry/118156/?skin=smt.

Cisco's network growth forecasts 2017-2022. Retrieved from https://www.cisco.com/c/en/us/solutions/collateral/service-provider/visual-networking-index-vni/white-paper-c11-741490.html

Columbus, L. (2018 April 15), The Future Of Manufacturing Technologies, 2018 [Blog post], Retrieved from https://www.forbes.com/sites/louiscolumbus/2018/04/15/the-future-of-manufacturing-technologies-2018/#771867a32995.

Desai, N. (2016 April 27), IT vs. OT for the Industrial Internet – Two Sides of the Same Coin? [Blog post], Retrieved from https://www.globalsign.com/en/blog/it-vs-ot-industrial-internet.

DeCesare, J. (2019 June 5) Why Organizational Readiness Assessments are Important [Blog pot], Retrieved from https://www.ispartnersllc.com/blog/why-organizational-readiness-assessments-are-important.

Deloitte Analysts et. Al (©2018) Exponential technologies in manufacturing [Online white paper], https://www.compete.org/storage/reports/exponential_technologies_2018_study.pdf.

Desai, N. (2016 August 25), IT vs. OT for the Industrial Internet – Two Sides of the Same Coin?," Industrial Internet Consortium, [Blog post], Retrieved from https://blog.iiconsortium.org/2016/08/it-vs-ot-for-the-industrial-internet-two-sides-of-the-same-coin.html.

DIN (2016 January 26) Updated German Standardization Roadmap 4.0 [White paper], Retrieved from https://www.din.de/en/din-and-our-partners/press/press-releases/updated-german-standardisation-roadmap-on-industry-4-0-110576.

Economist of London (2012 April 12),The third industrial revolution [Online article], Retrieved from https://www.economist.com/leaders/2012/04/21/the-third-industrial-revolution.

Economic and Social Commission for Asia and the Pacific (2018 October), Enhancing Cybersecurity for Industry 4.0 in Asia and the Pacific [White paper], Retrieved from https://www.unescap.org/sites/default/files/Cybersecurity_WorkingPaper-edit.pdf.

EDB Singapore (No date), The Singapore Smart Industry Readiness Index: Catalysing the transformation of manufacturing [White paper], Retrieved from https://www.gov.sg/~/sgpcmedia/media_releases/edb/press_release/P-20171113-1/attachment/The%20Singapore%20Smart%20Industry%20Readiness%20Index%20-%20Whitepaper_final.pdf.

Engelman, R. (© 2019), The Second Industrial Revolution 1870-1914 [Online article], Retrieved from http://ushistoryscene.com/article/second-industrial-revolution.

Executive Office of the President (2012 July) Report to the President on Capturing Domestic Competitive Advantage in Advanced Manufacturing [White paper], Retrieved from https://obamawhitehouse.archives.gov/sites/default/files/microsites/ostp/pcast_amp_steering_committee_report_final_july_17_2012.pdf.

Executive Office of the President (2014 October), Report to the President Accelerating U.S. Advanced Manufacturing [Government report], Retrieved from https://obamawhitehouse.archives.gov/sites/default/files/microsites/ostp/PCAST/amp20_report_final.pdf.

Flexis AG (© 2019) The Ultimate Guide to Industry 4.0 [Online article], Retrieved from https://www.flexis.com/en/solutions/industry-4-0.

Gaskell, A. (2018 September 17), Making Sense Of, And Making Progress With, Industry 4.0 [Online article], Retrieved from https://www.forbes.com/sites/adigaskell/2018/09/17/making-sense-of-and-making-progress-with-industry-4-0/#554cb4317ead

Gates, D. (©2018), Look out for i4.0 government incentives - KPMG Global [Blog post], Retrieved from https://home.kpmg/xx/en/home/insights/2018/10/look-out-for-i4-0-government-incentives.html.

Geissbauer, R. et. al. (2018 July 26), Digital Champions [Blog post], Retrieved from https://www.strategy-business.com/feature/Digital-Champions.

Geissbauer, R., Vedsø, J. and Schrauf, S. (2016 May 9), A Strategist's Guide to Industry 4.0 [Online article], Retrieved from https://www.strategy-business.com/article/A-Strategists-Guide-to-Industry-4.0.

Gold, S. (2018 January 9), Four Digital Trends Manufacturers Should Watch for in 2018[Blog post], retrieved from https://www.industryweek.com/technology-and-iiot/four-digital-trends-manufacturers-should-watch-2018.

Government of Japan Cabinet Office, (No date), Society 5.0 Backgrounder [Government report], Retrieved from https://www8.cao.go.jp/cstp/english/society5_0/index.html.

Granrath, L. (2017 August 29), Japan's Society 5.0: Going Beyond Industry 4.0 [Online article] retrieved from https://www.japanindustrynews.com/2017/08/japans-society-5-0-going-beyond-industry-4-0.

Green, T. (© 20219) This American's job? Make Singapore the world's first 'smart nation' [Blog post], Retrieved from https://www.hottopics.ht/9059/singapore-first-smart-nation/

Greenfield, D (2014 July 3), Industry 4.0 and OPC UA [Blog post], Retrieved from https://www.automationworld.com/industry-40-and-opc-ua.

Grenacher, M. (2018 April 11), Industry 4.0, The Smart Factory And Machines-As-A-Service [Online article,]Retrieved from https://www.forbes.com/sites/forbestechcouncil/2018/04/11/industry-4-0-the-smart-factory-and-machines-as-a-service/#4bac3be81dff.

Hamidi, S.R. et. al (2018 March 14), SMEs Maturity Model Assessment of IR4.0 Digital Transformation [Journal article abstract], Retrieved from https://link.springer.com/chapter/10.1007/978-981-10-8612-0_75.

Hanson-Muse, N. (2018 December 7), U.S. Companies Can Grow Their Business in Southeast Asia Thanks to New Agreements with Singapore [Blog post], Retrieved from https://blog.trade.gov/2018/12/07/u-s-companies-can-grow-their-business-in-southeast-asia-thanks-to-new-agreements-with-singapore.

Haseeb, A. (2018 January 10) "Higher education in the era of Industry 4.0 [Blog post], Retrieved from https://www.nst.com.my/education/2018/01/323591/higher-education-era-ir-40.

Hiskey, T. (2017 December 26), Three things industry 4.0 will change about how you work [Blog post], Retrieved from https://www.itproportal.com/features/three-things-industry-40-will-change-about-how-you-work.

Holt, V. (2018 October 6), Bringing Agile Concepts into the Physical Product Development World [Blog post], Retrieved from https://www.manufacturingleadershipcouncil.com/2018/10/06/bringing-agile-concepts-into-the-physical-product-development-world.

Hopewell, K. (2018 May 3), What is Made in China 2025? — and why is it a threat to Trump's trade goals? [Newspaper article], Retrieved from https://www.washingtonpost.com/news/monkey-cage/wp/2018/05/03/what-is-made-in-china-2025-and-why-is-it-a-threat-to-trumps-trade-goals.

Hoppe, S. (2017 June 19), There Is No Industry 4.0 without OPC UA [Blog post], Retrieved from https://www.automation.com/automation-news/article/there-is-no-industry-40-without-opc-ua.

Hughes, A. (2018 March 15), Standards in the IIoT: ISA-95 and Beyond [Blog post], Regtrieved from https://blog.lnsresearch.com/standards-in-iiot-isa-95-and-beyond.

Huifeng, H. (2018 July 18), Beijing did a tech reality check on its industrial champions [Newspaper article], Retrieved from https://www.scmp.com/news/china/economy/article/2155862/beijing-did-tech-reality-check-its-industrial-champions-results.

I4MS EU (2018 July 30), Industry 4.0, a new revolution in the manufacturing value chain [Blog post], Retrieved from https://medium.com/@i4ms_eu/industry-4-0-a-new-revolution-in-the-manufacturing-value-chain-382443f33f90.

Iansiti, M. (2015 June 30) ,The History and Future of Operations [Harvard Business Review article], Retrieved from https://hbr.org/2015/06/the-history-and-future-of-operations.

Institute for Security and Development Policy (2018 June), Made in China 2025 Backgrounder [Government report], Retrieved rom http://isdp.eu/content/uploads/2018/06/Made-in-China-Backgrounder.pdf.

Institution of Mechanical Engineers, (2019 April 23), Why the human factor is key to successful roll-out of Industry 4.0 tech [Blog post], Retrieved rom https://www.imeche.org/news/news-article/why-the-human-factor-is-key-to-successful-roll-out-of-industry-4.0-tech.

i-Scoop (No date), Gaps in Industry 4.0 readiness contribute to Industrie 4.0 Maturity Index [Blog post], Retrieved from https://www.i-scoop.eu/industry-4-0/gaps-industrie-4-0-maturity-index.

Katz, Sapper, and Miller (2015 February 23), The Third Industrial Revolution: Transforming Manufacturing [Blog post], Retrieved from https://www.ksmcpa.com/blog/the-third-industrial-revolution-transforming-manufacturing.

Kho, G. (2018 October 11) Singapore and Indonesia cooperate on Industry 4.0 solutions [Blog post], retrieved from https://www.opengovasia.com/singapore-and-indonesia-cooperate-on-industry-4-0-solutions.

Kibaroglu, O. (No date), Top 15 Innovations of the Fourth Industrial Revolution [Blog post], Retrieved from https://richtopia.com/emerging-technologies/industry-4-0-southeast-asia.

Learn Industrial Automation (© 2019), Open Platform Communications (OPC) [Online course description and video], Retrieved from https://www.lynda.com/Software-Development-tutorials/OPC/661768/713365-4.html.

Lichtblau, K. et. al. (2015 October), Industrie 4.0 Readiness [White paper], Retrieved from https://industrie40.vdma.org/documents/4214230/26342484/Industrie_40_Readiness_Study_1529498007918.pdf/0b5fd521-9ee2-2de0-f377-93bdd01ed1c8.

Limviphuwat, P. (2019 January 15), Asean industrial 4.0 plan first priority for Thailand [Blog post], Retrieved from https://elevenmyanmar.com/news/asean-industrial-40-plan-first-priority-for-thailand-asianewsnetwork.

Lim, J. (2018 October 25), Malaysia to introduce the National Industry 4.0 Policy Framework [Blog post], Retrieved from https://www.theedgemarkets.com/article/malaysia-introduce-national-industry-40-policy-framework.

Lydon, W. (2018 June 18), How Industry 4.0 and Digitization Improves Manufacturing Responsiveness, Quality and Efficiency [Blog post], retrieved from https://automation.isa.org/industry-40-digitization-improve-manufacturing-responsiveness-quality-efficiency-iot.

Lydon, W. (2017 June 1), Industry 4.0 for process [Blog post], Retrieved from https://www.isa.org/intech/20170601.

Malaysia Productivity Corporation (© 2018), The Race Towards Industry 4.0 [White paper], Retrieved from http://www.mpc.gov.my/wp-content/uploads/2018/11/The-Race-Towards-Industry-4.0.pdf.

Masson, C. (2018 April 11) Why the OPC UA Standard – and What's Next? [Blog post], Retrieved from https://cloudblogs.microsoft.com/industry-blog/manufacturing/2018/04/11/why-the-opc-ua-standard-and-whats-next.

McKeon, A. (2018 July 5), Get to Industry 4.0 with a smart factory roadmap [Blog post], Retrieved from https://searcherp.techtarget.com/feature/Get-to-Industry-40-with-a-smart-factory-roadmap.

McKinsey Global Institute (2018 May), Skill Shift: Automation and the Future of the Workforce [White paper], Retrieved from https://www.mckinsey.com/~/media/McKinsey/Featured%20Insights/Future%20of%20Organizations/Skill%20shift%20Automation%20and%20the%20future%20of%20the%20workforce/MGI-Skill-Shift-Automation-and-future-of-the-workforce-May-2018.ashx.

Meering, C. (2018 July 5), Interoperability is key to boosting Industry 4.0 - IoT Agenda [Blog post], Retrieved from https://internetofthingsagenda.techtarget.com/blog/IoT-Agenda/Interoperability-is-key-to-boosting-Industry-40.

Min, C.Y. (2018 January 13), Singapore well positioned to gain from Industry 4.0 [Online article], Retrieved from https://www.straitstimes.com/business/economy/singapore-well-positioned-to-gain-from-industry-40.

Mittal, S. et. al. (2018 October 5), A Critical Review of Smart Manufacturing & Industry 4.0 Maturity Models: Implications for Small and Medium-sized Enterprises (SMEs) [Blog post], Retrieved from https://www.researchgate.net/publication/328726147_A_Critical_Review_of_Smart_Manufacturing_Industry_40_Maturity_Models_Implications_for_Small_and_Medium-sized_Enterprises_SMEs.

MNBB Studio, Industry 4.0 and the fourth industrial revolution: guide to Industrie 4.0 [Online article] Retrieved from https://www.i-scoop.eu/industry-4-0.

Mohammed M. Mabkhot et. al. (2018 1 June), Requirements of the Smart Factory System: A Survey and Perspective [Online Article], Retrieved from https://www.mdpi.com/2075-1702/6/2/23/pdf

Moore, J. (2015 22 May), IoT will force CIOs to enter the realm of operational technology [Blog post], Retrieved from https://internetofthingsagenda.techtarget.com/news/4500246872/IoT-will-force-CIOs-to-enter-the-realm-of-operational-technology.

Mosch, C. (© 2019), Industrie 4.0 Readiness study – VDMA [Online article], Retrieved from https://industrie40.vdma.org/en/viewer/-/v2article/render/15525817.

Moyne, J. and Iskandar, J. (2017 July 12), Big Data Analytics for Smart Manufacturing: Case Studies in Semiconductor Manufacturing [Research paper], Retrieved from file:///C:/Users/patru/AppData/Local/Packages/Microsoft.MicrosoftEdge_8wekyb3d8bbwe/TempState/Downloads/processes-05-00039%20(2).pdf.

NAMRI/SME (2014 July), Advanced Manufacturing Initiatives: A National Imperative [White paper], Retrieved from https://wir-en.s3.amazonaws.com/wp-content/uploads/2014/07/NAMRI_White_Paper.pdf.

Nayan, A. (no date), I4.0: Introduction to Concepts & 11 Technology Pillars [Downloadable brochure]. Retrieved from http://www.psdc.org.my/assets/documents/training/industry-40/i40-introduction-to-concepts-technology-pillars.pdf.

Newton Fund (2018 January 18), Newton Fund Industry-academia Partnership Programme – Thailand [Blog post], Retrieved from https://euraxess.ec.europa.eu/worldwide/asean/newton-fund-industry-academia-partnership-programme-thailand.

Oleśków-Szłapka, J. and Stachowiak, A. (2019 January), The Framework of Logistics 4.0 Maturity Model [Conference paper], Retrieved from https://www.researchgate.net/publication/326749347_The_Framework_of_Logistics_40_Maturity_Model.

OPC Foundation (No date), OPC Unified Architecture Interoperability for Industrie 4.0 and the Internet of Things [Online white paper], Retrieved from https://opcfoundation.org/wp-content/uploads/2016/05/OPC-UA-Interoperability-For-Industrie4-and-IoT-EN-v5.pdf.

Opensignal. The State of Mobile Network Experience. Southeast Asia Report. Retrieved from https://www.opensignal.com/reports/2019/05/global-state-of-the-mobile-network

Ostdick, N. (2017January 14) How Industry 4.0 is Changing Supply Chain Management [Blog post], Retrieved from https://blog.flexis.com/how-industry-4.0-has-changed-global-supply-chain-management.

Oxford Business Group (© 2019) Industry 4.0 stands to disrupt existing industrial models in emerging economies [Online article], Retrieved from https://oxfordbusinessgroup.com/overview/high-gear-economies-around-world-are-preparing-opportunities-and-challenges-brought-about-next.

Oxford Economics and Cisco (2018 September), Technology and the future of ASEAN jobs: The impact of AI on workers in ASEAN's six largest economies [White paper], Retrieved from https://www.cisco.com/c/dam/global/en_sg/assets/csr/pdf/technology-and-the-future-of-asean-jobs.pdf.

Pelizzo, G. (2016 June 22), Industry 4.0: A Concrete Roadmap [Blog post], Retrieved from https://www.techedgegroup.com/blog/industry40-a-concrete-roadmap.

Phoenix, J. (2018 December 07), Top 10 manufacturing trends for 2019 [Blog post], Retrieved from https://www.manufacturingglobal.com/top-10/top-10-manufacturing-trends-2019.

Pilgrim Quality Solutions (© 2019), Smart Quality Management: Impact of Industry 4.0 on QMS [e-book], Retrieved from http://info.pilgrimquality.com/smart-quality-management-ebook.

Price, N. (2015 October 30), Moving Towards An Outcome Economy [Blog post], Retrieved from https://www.smartindustry.com/blog/smart-industry-connect/moving-towards-an-outcome-economy.

Principia Scientific. Wireless Radiation: Stop The 5G Network On Earth And In Space, Devastating Impacts On Health And The Environment. Petition. Retrieved from https://principia-scientific.org/petition-26000-scientists-oppose-5g-roll-out/

Rifkin, J. (2016 January 14) The 2016 World Economic Forum Misfires with its Fourth Industrial Revolution Theme [Online article], Retrieved from https://www.huffpost.com/entry/the-2016-world-economic-f_b_8975326.

Rizk, A. (2016 December 7), Industry 4.0: the urgency of data standardisation [Blog post], Retrieved from https://www.manufacturingglobal.com/lean-manufacturing/industry-40-urgency-data-standardisation.

Rodriguez, K. (2016 May 12), Have we reached the fourth industrial revolution? [Blog post], Retrieved from https://execed.economist.com/blog/industry-trends/have-we-reached-fourth-industrial-revolution.

Salesforce.com (©2019), Meet the Three Industrial Revolutions [Online article], Retrieved from https://trailhead.salesforce.com/en/content/learn/modules/learn-about-the-fourth-industrial-revolution/meet-the-three-industrial-revolutions.

Santander Trade Portal (© 2019) Indonesia: Foreign Investment [Online data sheet], Retrieved from https://en.portal.santandertrade.com/establish-overseas/indonesia/foreign-investment.

Santoso, S. (2017 May 10), How can ASEAN nations unlock the benefits of the Fourth Industrial Revolution?   [Online article], Retrieved from https://www.weforum.org/agenda/2017/05/how-can-asean-nations-unlock-the-benefits-of-the-fourth-industrial-revolution.

Schmarzo, J. (2018 January 8), Manufacturing, the Smile Curve and Digital Transformation [Blog post], Retrieved from https://infocus.dellemc.com/william_schmarzo/manufacturing-the-smile-curve-and-digital-transformation.

Schweichhart, K. (2016 April), Reference Architectural Model Industrie 4.0 (RAMI 4.0) [Online white paper], Retrieved from https://ec.europa.eu/futurium/en/system/files/ged/a2-schweichhart-reference_architectural_model_industrie_4.0_rami_4.0.pdf.

Singapore Economic Development Board (2019 March 22), The Singapore Smart Industry Readiness Index [Blog post], Retrieved from https://www.edb.gov.sg/en/news-and-events/news/advanced-manufacturing-release.html.

Singaporean-German Chamber of Industry and Commerce (2016 September) Future of Manufacturing [White paper], Retrieved from https://www.sgc.org.sg/fileadmin/AHK_Singapur/PUBLICATIONS/Future_of_Manufacturing.pdf.

Singapore Government's Second Public Consultation on 5G Mobile Services and Networks. Retrieved from https://www2.imda.gov.sg/regulations-and-licensing/Regulations/consultations/Consultation-Papers/2019/Second-Public-Consultation-on-5G-Mobile-Services-and-Networks

Smart Nation and Digital Government Office (© 2018), Smart Nation: The Way Forward Executive Summary [White paper], Retrieved from https://www.smartnation.sg/docs/default-source/default-document-library/smart-nation-strategy_nov2018.pdf.

Smart Nation Singapore (© 2019) Smart Nation Singapore [Videos], Retrieved from https://www.smartnation.sg.

Smith, Nigel (2018 November 18), Industry 4.0: Dispelling myths surrounding smart manufacturing [Blog post] Retrieved from https://www.automation.com/automation-news/article/industry-40-dispelling-the-myths-surrounding-smart-manufacturing.

Sontag, D. (2018 June 28), to rock Industry 4.0? These 6 challenges are in your way [Blog post], Retrieved from https://medium.com/the-industry-4-0-blog/want-to-rock-industry-4-0-these-6-challenges-are-in-your-way-d9d841ebc51d.

Spiropoulos, E. (2017 February 3), A look at IIoT and the flattening of ISA-95 Industrial Automation Architecture models [Blog post], Retrieved from https://www.linkedin.com/pulse/look-iiot-flattening-isa-95-industrial-automation-eugene-spiropoulos.
Tan, A. (2018 October 31), Inside Schneider Electric's Batam smart factory [Blog post], Retrieved from https://www.computerweekly.com/news/252451662/Inside-Schneider-Electrics-Batam-smart-factory.

Tan, Z. Y. (2018 July 13) Industry 4.0: The journey towards automation [Blog post], Retrieved from https://www.theedgemarkets.com/article/industry-40-journey-towards-automation

Tarver, E. (2019 April 19), Value Chain vs. Supply Chain: What's the Difference? [Online article], Retrieved from https://www.investopedia.com/ask/answers/043015/what-difference-between-value-chain-and-supply-chain.asp.

Tee, J. (November 13), Key to manufacturing success in ASEAN [Blog post], Retrieved from https://www.businesstimes.com.sg/hub/asean-singapore-2018/key-to-manufacturing-success-in-asean.

Thailand Board of Investment (2017 June 15), Thailand's Automation and Robotics: The Rise of Automation and Robotics Industries [White paper], Retrieved from https://www.boi.go.th/upload/content/BOI-brochure%202016-automation-20170615_14073.pdf.

The Atlantic, (©2017), Powering the Smart Factory with the Internet of Things [Online report], Retrieved from https://www.theatlantic.com/sponsored/vmware-2017/iot-manufacturing/1751/.

The Guardian. Mobile phones and cancer: the full picture. Retrieved from https://www.theguardian.com/technology/2018/jul/21/mobile-phones-are-not-a-health-hazard

The Manufacturer (2018 July 22), Industry 4.0: A sustainable roadmap for midsized manufacturers [Blog post], Retrieved from https://www.themanufacturer.com/articles/industry-4-0-a-sustainable-roadmap-for-midsized-manufacturers

The Straits Times (2019 February 11), What Singapore firms want the most from Budget 2019 is funding for tech adoption, says SBF poll [Blog post], Retrieved from https://www.straitstimes.com/business/economy/what-singapore-firms-want-the-most-from-budget-2019-is-funding-for-tech-adoption

The Worldfolio (2018 October), Education pioneers talent development & industry 4.0 economy [Blog post], Retrieved from http://www.theworldfolio.com/news/education-pioneers-talent-development--industry-40-economy/4387.

Togue, M. (2018 August 18), 6 steps for implementing Industry 4.0 [Blog post], Retrieved from https://www.lanner.com/en-us/insights/blog/6-steps-to-industry-4-0-success.html.

Totton, R. (2016 May 30) Why is Singapore struggling with digital adoption? [Online article], Retrieved from https://www.weforum.org/agenda/2016/05/why-is-singapore-struggling-with-digital-adoption.

Towers-Clark, C. (2019 February 20), Big Data, AI & IoT Part Two: Driving Industry 4.0 One Step At A Time [Online article], Retrieved from https://www.forbes.com/sites/charlestowersclark/2019/02/20/big-data-ai-iot-part-two-driving-industry-4-0-one-step-at-a-time/#261c24b323a0.

Traitrong, W. (2018 October 23), Singapore on track to pursue industry 4.0 vision[Blog post], Retrieved from https://www.nationthailand.com/Economy/30357021.

United Overseas Bank, Ltd. (2018 October 22), UOB boosts its support of SMEs through innovation and internationalisation programme [Press release], Retrieved from UOB-SMEs-innovation-and-internationalisation-programme.pdf.

United States Department of Homeland Security (2018 May 8), Global Industry 4.0 Market & Technologies 2018-2023 [Online report summary], Retrieved from https://www.globenewswire.com/news-release/2018/05/08/1498224/0/en/Global-Industry-4-0-Market-Technologies-2018-2023-Market-is-Expected-to-Reach-1-Trillion-by-the-Early-2030s.html.

United States National Science and Technology Council (2018 October), Strategy for American Leadership in Advanced Manufacturing [Government report], Retrieved from https://www.whitehouse.gov/wp-content/uploads/2018/10/Advanced-Manufacturing-Strategic-Plan-2018.pdf.

University of Warwick (no date), An Industry 4 readiness assessment tool [White paper]. Retrieved from https://www.crimsonandco.com/wp-content/uploads/2017/10/Industry-4-readiness-assessment-tool-report-Oct-2017.pdf.

Vardi, M.Y. (2018 September 1), What the Industrial Revolution really tells us about the future of automation and work [Online article], Retrieved from http://theconversation.com/what-the-industrial-revolution-really-tells-us-about-the-future-of-automation-and-work-82051.

Vietnam Investment Review (2018 August 30) IT key to Vietnam's Industry 4.0 future [Blog post], Retrieved from https://www.vir.com.vn/it-key-to-vietnams-industry-40-future-62095.html.

Vietnam Investment Review (2018 November 9), WEF drives Vietnam to an Industry 4.0 future [Blog post], Retrieved from https://www.vir.com.vn/wef-drives-vietnam-to-an-industry-40-future-62296.html.

World Economic Forum (2016 May 31), South-East Asia's digital jobs revolution – in 5 charts [Blog post], Retrieved from https://www.weforum.org/agenda/2016/05/south-east-asia-digital-jobs-5-charts.

World Economic Forum (© 2018), Readiness for the Future of Production Report 2018 WEF + A.T. Kearney Date [Online report], Retrieved from http://www3.weforum.org/docs/FOP_Readiness_Report_2018.pdf.

World Economic Forum (2019 January), Fourth Industrial Revolution Beacons of Technology and Innovation in Manufacturing [White paper], Retrieved from http://www3.weforum.org/docs/WEF_4IR_Beacons_of_Technology_and_Innovation_in_Manufacturing_report_2019.pdf.

World Economic Forum (no date), Shaping ASEAN's Future Readiness [Online article], Retrieved from https://www.atkearney.com/web/world-economic-forum/article/?/a/shaping-asean-s-future-readiness.

Yeo, D. (2017 May 7), Industry 4.0: Interoperability What is it and Why Should You Care? [Blog post], Retrieved from http://supplychainasia.org/industry-4-0-interoperability-care.

Zippel, S. (2017 August 8), How to implement Industry 4.0 within the Process Industry [Blog post], Retrieved from https://medium.com/marcus-evans-webinars/how-to-implement-industry-4-0-within-the-process-industry-by-stefan-zippel-1616381aa74d.

Copyright@2021 by Michael Deng & Colin Koh

ISBN 978-1-7105-213-368

www.asean4ir.com